Designing the

NEW GENERATION John Deere Tractors

Merle L. Miller

Published by the
American Society of Agricultural Engineers
2950 Niles Road
St. Joseph, Michigan

About ASAE — the Society for engineering in agricultural, food, and biological systems

ASAE is a technical and professional organization of members committed to improving agriculture through engineering. Many of our 8,000 members in the United Stated, Canada, and more than 100 other countries are engineering professionals actively involved in designing the farm equipment that continues to help the world's farmers feed the growing population. We're proud of the triumphs of the agriculture and equipment industry. ASAE is dedicated to preserving the record of this progress for others. This book joins many other popular ASAE titles in recording the exciting developments in agricultural history.

Designing the New Generation John Deere Tractors
Editor and Book Designer: Melissa Carpenter Miller

Library of Congress Catalog Card Number (LCCN): 99-72167
International Standard Book Number (ISBN): 1-892769-04-2

Printed in the U.S.A.

Acknowledgments

As in all books of this type, considerable confirmation of facts and experiences is necessary. The author is indebted to many persons and wishes to acknowledge the valuable assistance they provided. The 1959 Seminar Booklets (not publicly available) were the greatest help in compiling this book. Illustrations, highlights, and direct quotations from these booklets are used.

Topics and respective authors of these booklets were:

General – Design Objectives, Tractor Arrangement, Weight, Power, Models Etc. – Wallace DuShane (deceased), Russell Candee (deceased)
General – Controls, Operator Station, Etc. – J. Daniel Gleeson, Robert Woolf
Diesel Engines – Dean Rudig, John Townsend (deceased)
Spark Ignition Engines – John Sandoval, Robert Hall (deceased)
Transmission, Final Drive, and PTO – Vernon Rugen, Richard Fox (deceased), Richard Doerfer
Brakes and Power Steering – Richard Wittren, Harlan Jensen
Implement Hitch – Christian Hess (deceased), Wendell Van Syoc
Hydraulic System – General, Pump, Valving, Auxiliary Functions, Rockshaft – Edward Fletcher, Ober Smith (deceased), Martin Borchelt

A number of persons graciously reviewed and commented on the various chapters in this book:

General Design Objectives – J. Daniel Gleeson
Spark Ignition Engines – John Sandoval
Diesel Engines – Sydney Olsen (deceased)
Transmission, Final Drive and PTO – Vernon Rugen
Hydraulic System, Brakes and Steering – Edward Fletcher
Implement Hitch – Harold Kienzle
Operator Station, Controls and Chassis – J. Daniel Gleeson
Development and Testing – Wilbur Davis (deceased), Kenneth Murphy, Wendell Van Syoc
Service – Fred Hileman
Manufacturing and Reliability – E. Dean Ethington

Wilbur Davis was especially helpful in encouraging the creation and completion of this account.

Others who have contributed in various ways are: W.H.F. Purcell (retired) and Donald M. Genero of Henry Dreyfuss Associates, Dr. Leslie Stegh of John Deere Archives, Robert Avery (retired) of John Waterloo Tractor Works Service Department, and Harold Brock (retired), James Kress (retired), Warren Wiele (retired), Harlan W. Van Gerpen (retired), and Alice McGrath of the Product Engineering Center. Also, Jack Cherry, Two-Cylinder Club and (retired) John Deere Waterloo Tractor Works Foundry.

The help of Donna Hull, Melissa Miller, and others from ASAE have changed and guided the book into a finished product.

And lastly, I am indebted and grateful to my wife, Doris Marie (Fritzie) for her support, help, and tolerance.

Dedication

This book is dedicated to the memory of my late good friend, Wilbur Davis. His encouragement and many contributions have stimulated its completion. His strong characteristics of integrity, forthrightness, persuasion, and personal productivity were substantial factors in the level of customer satisfaction with the field performance of these tractors.

About the Author

Merle Miller retired from Deere & Company in 1984 after 38 years with the company. He is a graduate of the University of Missouri-Columbia with a B.S. in Mechanical Engineering. He served in the military as Engineering Officer on LST ships in the Mediterranean and U.S. waters. He advanced to Lieutenant, USNR, before resigning his commission in 1953. Following active military service, he began his career with Deere & Co., rising to the title of senior division engineer..

During his years with Deere, Mr. Miller worked as supervisor on testing and design groups. Some of the projects he worked on included the 8010 tractor, the 1020 and 2020 tractors, roll-over protection, Deere's own enclosed cab, the 6000 and 7000 series tractors, and several series of the four-wheel-drive tractors.

Mr. Miller has been active as a director of the Council of Agricultural Science and Technology (CAST) and served on several American Society of Agricultural Engineers (ASAE) boards and committees. He is an ASAE Fellow Member.

Introduction

The atmosphere in the Dallas Coliseum was electric that August day in 1960. Shiny green and yellow tractors were being revealed before an awed audience of about 6,000 John Deere dealer personnel, company officials, and the press. They came from all over the United States and Canada, as well as other foreign countries. Most were not prepared to see the extent of product change for which they had been flown to Dallas. It was exciting and stimulating to be part of a marketing organization that dared to provide these innovative and advanced product designs — truly a "New Generation."

The events and the conditions that guided the decisions leading up to this event were fortunate in their timing. These circumstances included conditions within Deere & Company and the changing marketplace of power farming in the United States. Later events, plus changes in the competitive products, only emphasize how well-suited these products were for their time.

Seldom has a well-worn product near the end of its useful life been changed to one that creates a new product so completely ahead of the competition. This scenario resulted in competitors later evolving similar advantages in their products.

The "New Generation" tractors included four models in order of increasing power: 1010, 2010, 3010, and 4010. The 1010 was principally a new four-cylinder engine in the 430 tractor. The 2010 tractor was mostly new parts and represented a new larger size for the Dubuque factory. The 3010 and 4010 were completely new designs made in the Waterloo factory.

How the "New Generation" 3010 and 4010 tractors came into being is the story told within these pages. The size of the program was recognized by senior management to require a substantial cost investment. This recognition resulted in help from all levels of management, and from all disciplines. (The more complete redesigns of the 1010 and 2010 tractors were introduced in 1965 as the 1020 and 2020. They were also built in Dubuque. The development of these first Deere "worldwide" designs, is another important John Deere engineering and marketing story. That story, not told here, includes their manufacture in foreign factories, and expanding to more power sizes.)

Most engineering employees who were fortunate to work on

The four New Generation tractors

the 3010 and 4010 tractor project are now still living. Some persons involved with these tractors have now gone to their reward. Telling the story of this revolutionary change now is timely while the memories of those years of effort are still with us.

Most persons who worked on these new products have recently retired from the company. Many feel that they worked during a most exciting time in the tractor industry. Through this period, Deere & Company experienced significant growth in their business and increase in market share for their products. Those persons associated with this new tractor program believe they had the fortunate experience of a lifetime.

This cutaway shows the inner workings of a 4010 diesel tractor.

Not enough can be said about the leadership and motivation provided by Deere & Company. The managers were encouraged to be innovative, to be expeditious in their activities, and to delegate responsibility to the working level.

This book, by a retired John Deere engineer about a major engineering program, contains some engineering design terminology unique to farm tractors. Much of the background information used by the author is taken from eight booklets prepared in 1959 for seminars informing the factory supervisors and the current products engineers. The Program Manager at that time was the late W. (Wally) H. DuShane who introduced the series to the latter group with these remarks:

> *"You men of the Waterloo Tractor Works (Current Products) Engineering Department are about to assume the full responsibility of the design of a new series of tractors. In so doing you will be called upon to answer many questions as to why the tractor has been so designed. Some of these questions will not be easy to answer. To others there may not be an answer - other than that was the only way we knew how to design it. But for the most of the questions there is a real reason for arriving at the particular design."*

The objective of this book is to give some of these reasons by acquainting the prospective readers with the history of what, how, and why this program happened. Chapters 1, 2, and 3 describe the conditions and events leading to these new products. Chapters 4 through 8 explain the design choices made in each of the major tractor groups. Chapters 9, 10, and 11 describe the steps necessary to change the lines on the drawing board into steel, iron, and rubber rolling off the assembly line to customers. Chapter 12 describes some of the marketing introduction and security provisions. The last chapter shows the final product.

Any explanation or description of a successful product program can read like a love affair between the topic and the author. So is this book, though the author's participation in the principal product effort was only on the fringes. (When a new product succeeds, everyone claims involvement, but when the product change has problems few publicize their connection.) The author was aware of most events as they occurred and knew most of the persons involved in the product design and development.

Table of Contents

The Company and Its Support

The development of the multi-cylinder tractor design at Deere & Company in the 1950s was due not just to the changes in tractor power requirements. Nor was it simply the need for a modern appearing family of tractors. Or even because of the condition of the factory and its design and production equipment, which will be discussed later in this book. Of greater importance, however, was the growing sense of need among the central company's managers. They recognized the need for tractors to continue to offer increasing value to the customer, and they became convinced it would not happen with the old two-cylinder design. Without the encouragement of top corporate management, any design could fail to become a marketable product.

"I will never put my name on an implement that doesn't have in it the best that is in me." — John Deere ("Makers of Plenty – John Deere, The Man and the Company," Deere & Co.)

John Deere's Influence on Company Philosophy

Deere & Company's support philosophy really started with John Deere himself. Since others have thoroughly recounted his history and the early company beginnings, these subjects do not require complete restating here. However, his character and personal business objectives had a strong influence in succeeding years. These qualities helped shape the environment that much later produced the New Generation tractors.

His characteristics that had lasting effect are stated in part by Wayne G. Broehl Jr., in *John Deere's Company*, ". . . he did have a knack for organization, an abiding concern for quality, and a feeling for the role of the agricultural equipment industry in America's growth." John Deere's early concern for quality is noted by his statement, "I will never put my name on an implement that doesn't have in it the best that is in me." This responsible attitude in producing a quality product has continued throughout the years. Some of the trademarks stressed quality. However, this attitude was much more than just an advertising tool. For example, recently sophisticated prediction techniques have provided predesign reliability estimates. Also, testing of John Deere experimental tractors has been extensively executed to uncover the prospect of premature failure in the production models. In addition, monitoring tractor production has attempted to limit the shipping of sub-standard units.

Deere & Company adopted this trademark in 1950 which emphasized quality.

Engineering judgments on reliability have not been based upon just "getting by," but rather upon whether the product change was right for the intended use and application. "Right" carries the meaning that the concept was one that decision and design makers could be proud of and live with comfortably. In this way, the expectation was that the customer's satisfaction would be achieved. This attitude, to a large degree, pervaded the managerial ranks and was passed along to subordinate associates.

John Deere's belief in the future role of the agricultural equipment industry helped shape the company. This belief gave his early company the staying power to continue ownership and successful management during his and his son Charles' years. He continuously sought improvements in his product line and tried various management and financing concepts. Deere believed in expanding his product line to have more to offer his customers (and, of course, enhance the company's profits).

Expansion of the product line allowed his successors to acquire other companies that had unique products to complete the Deere & Company line. Rather than be merely a marketing or manufacturing organization, he combined the two functions. The result became more than just the addition of the two. This concept continues to serve the company in recent years.

In the organizational arena, results speak louder than words. Small businesses, as they become large corporations, face two obstacles. First, the transition from the early, entrepreneurial operation to one in which delegation of responsibilities must happen. Second, bringing into the organization additional persons of capability who contribute in essential ways, and do not become mere "yes" men.

History shows that the Deere organization overcame these obstacles. Family-owned or managed corporations sometimes name as president a family relative who lacks the leadership, insight, aggressiveness, or financial acuity of his or her ancestors. The business then suffers. John Deere was fortunate in being followed by his son, Charles Deere. Charles was followed by other succeeding related family members; William Butterworth, Charles Wiman, and William Hewitt. Each brought the necessary leadership skills for his time and industry conditions.

In addition, the branch house marketing concept and the decentralized factory managements brought new responsible persons into the decision-making process. Many of these people progressed to higher positions in central Deere management. Their new capabilities strengthened the company in most instances. Some left to take positions of promise in other companies. Again, their vacancies were filled by other capable people.

John Deere, Deere & Company founder and president from 1869-1886.

A humorous story is told about the testing of an experimental John Deere plow. This plow had a new design of the steel beam connecting the plow bottom to the pulling attachment. The designers were so proud of their work that they placed the warranty label in the center of the new feature. In early tests, the field test supervisor telephoned the factory engineering office to report the plow beam failure. In response to the office's inquiry on the break location, he said: "right through the guarantee!"

Charles Deere, Deere & Company president 1886-1907.

William Butterworth, Deere & Company president 1907-1928.

Charles Deere Wiman, Deere & Company president 1928-1955.

William Hewitt, Deere & Company president 1955-1982.

This wide and varied experience, along with personal motivation, helped make the company sensitive to the changing needs of the customer. It also maintained good financial stability, modern manufacturing centers, and a strong marketing organization.

The Development of Deere's Tractor Line

During the 1910s, company management was aware of the growing need to add tractors to the product line. Several internal designs were investigated, and the purchase of successful tractor lines were considered. Several major competitors already had a tractor line, some having started with steam powered units.

The period between 1915 and 1925 saw many small manufacturers making a tractor product. Some tractors were reliable, others were not. The unreliable tractors created a great deal of disgust and concern among the farmers. Out of this situation arose the Nebraska Test Board and its test procedures. These tests made public the power and mechanical failures of the various makes. This practice assisted in the success of the most reliable makes of tractors.

Deere deliberated for several years over the need to enter the tractor business. Tractor availability from most key competitors threatened the sales of Deere implements, especially the bread-and-butter line of plows. Eventually, this forced management to act decisively.

In 1918, Deere purchased the Waterloo Gasoline Engine Company, a firm that had a tractor development history dating from 1892. This company, named after its location in Iowa, produced stationary engines and two Waterloo Boy tractor models. These models were the "R" and "N." The "N" model was later tested in 1921 at Nebraska and developed 25 belt horsepower.

This is a copy of the check used in 1918 to purchase the Waterloo Gas Engine Company.

DEERE & COMPANY
MOLINE

VOUCHER NO. G.3163

March 14

CHECK NO. G66905

VOUCHER IN FAVOR OF

Waterloo Savings Bank

$ 2,100,000.0

Two Million, One Hundred Thousand 00/100 --------------- DOLLARS

To the First National Bank, Chicago, Ill.

The Model D tractor was the first new design under Deere ownership.

Deere management supported continued tractor design improvement in the newly purchased factory's products. The two-cylinder Waterloo Boy was replaced in 1923 with its first John Deere designed tractor, the Model D. This original 27-belt horsepower model continued in production, with improvements, until 1953. This span of 30 years made the Model D the longest model run of any John Deere tractor production.

Major improvements during those years included power increases to 42 belt horsepower and sheet-metal styling. Rubber tires, power take-off shaft, electric starting, and lights became available. The design was comparatively simple, reliable, and durable. The tractors were also easy and economical to operate and service. The Model D became a popular tractor, especially in non-row-crop areas. Deere managerial leadership under President Charles D. Wiman (1928-1955) supported this continued product improvement in a more liberal manner than the previous William Butterworth (1907-1928) administration.

The support given in tractor improvement during the 1930s, 1940s, and 1950s was timely and enabled the addition of other models and sizes. These changes improved the tractor line by making it more suitable to meet the needs of additional customers whose tractor purchases were increasing. As part of the marketing strategy, the inherent simplicity and rugged appearance of the engine parts were stressed.

This illustration of a John Deere two-cylinder engine crankshaft was often used in Deere advertising to stress the inherent simplicity and rugged appearance of these engine parts.

These later tractor models were also of two-cylinder design. John Deere and two-cylinder tractors became synonymous. Additional tractor versions were provided to better meet the requirements of special or regional crops such as orchards and beets. Other early models available following the "D" were the "GP," "L," and "LA" tractors.

Many readers may be acquainted with the more recent model listings of "A," "B," "G," "GM," "H," "M," and "R" alphabet designations. They were replaced by the "40," "50," "60," "70," and "80" designations in the 1950s. The latter part of the same decade saw three digit model designations with the last zero replaced with "20," then still later with "30." With model number changes came other significant improvements, including power increases. These tractors all had two-cylinder engines, and all the Waterloo-built tractors were of horizontal cylinder arrangement with a crosswise crankshaft.

The Model M, and its successors, "40," "420," "430," and "435" were made in a new factory opened in 1946 in Dubuque, Iowa. This was an outgrowth of the organization that had produced the Models L and LA at the Deere Wagon Works in Moline. These tractors had two vertical in-line cylinders and a longitudinal (perpendicular to the rear axle) crankshaft.

Clouds on the Horizon

The decade of the 1950s was an eventful one for Deere & Company. In 1955, President Wiman died. His son-in-law, William A. Hewitt, was elected president of the company. The sales volume was steadily increasing with the general post-war prosperity. The expansion of U.S. farmer's markets was aided by the creation and execution of the Marshall Plan foreign food program. John Deere tractors increased in power and popularity thus contributing to Deere surpassing the previous number one manufacturer in agricultural equipment sales — International Harvester.

With all this success however, there were dark clouds on the horizon for Deere tractors. First, the decade saw strong demands for increased power in tractors. There was no obvious plateauing of this demand. A second cloud was the fact that the tractor market became primarily the replacement of existing tractors. This increased the competition among the surviving tractor manufacturers. Yet another cloud was the increased age of the factory equipment. Many factory machines had been used for

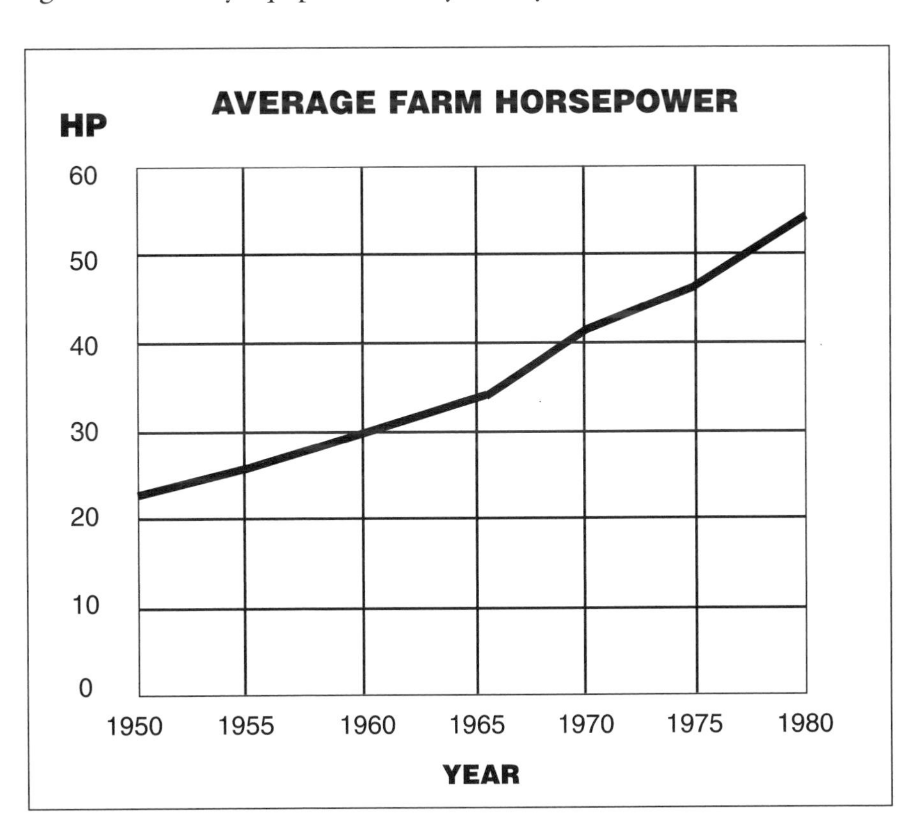

This chart depicts the trend toward increasing horsepower on farms beginning in 1950.

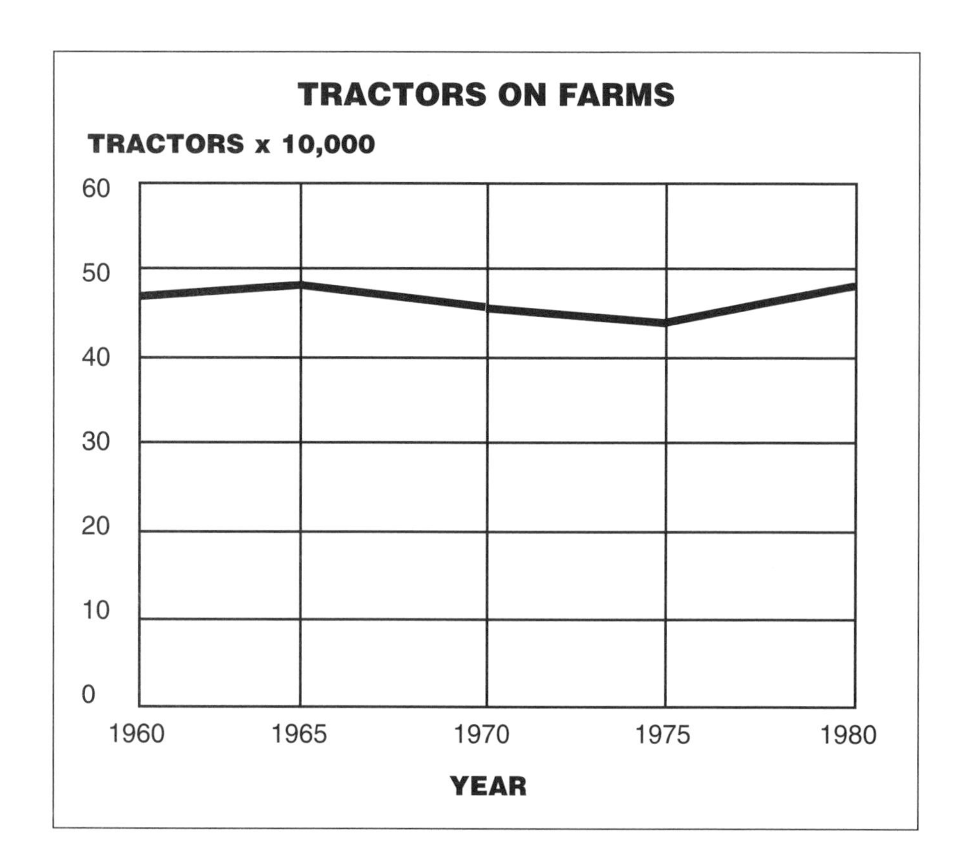

Tractor quantities on farms did not change greatly from 1960 to 1980. When compared with the illustration on page 5, the horsepower of the average tractor was increasing substantially, and the market for new tractors was largely for replacement.

numerous years, some for as long as 20 years. Many parts were made on critical equipment that was aged and worn.

Still another looming cloud was the increasing demand for operator comfort. This demand was partly met in 1954 with the first factory-equipped power steering option. This development followed earlier introductions of mechanical and hydraulic implement raising. Despite these steps, a tractor driving operation had its miserable moments. Few tractors were available with protection from elements of wind, rain, snow, sun, heat, and cold. The noise levels were reduced by exhaust mufflers, but were no comparison to contemporary cars and trucks. Improved seating was being provided, both in durability and comfort, though far less in comfort than in other types of vehicles.

This aerial view of the John Deere Waterloo Tractor Works was taken in 1958 — just before the New Generation tractors were made.

This illustration was used in a sales brochure which described the new power steering factory option.

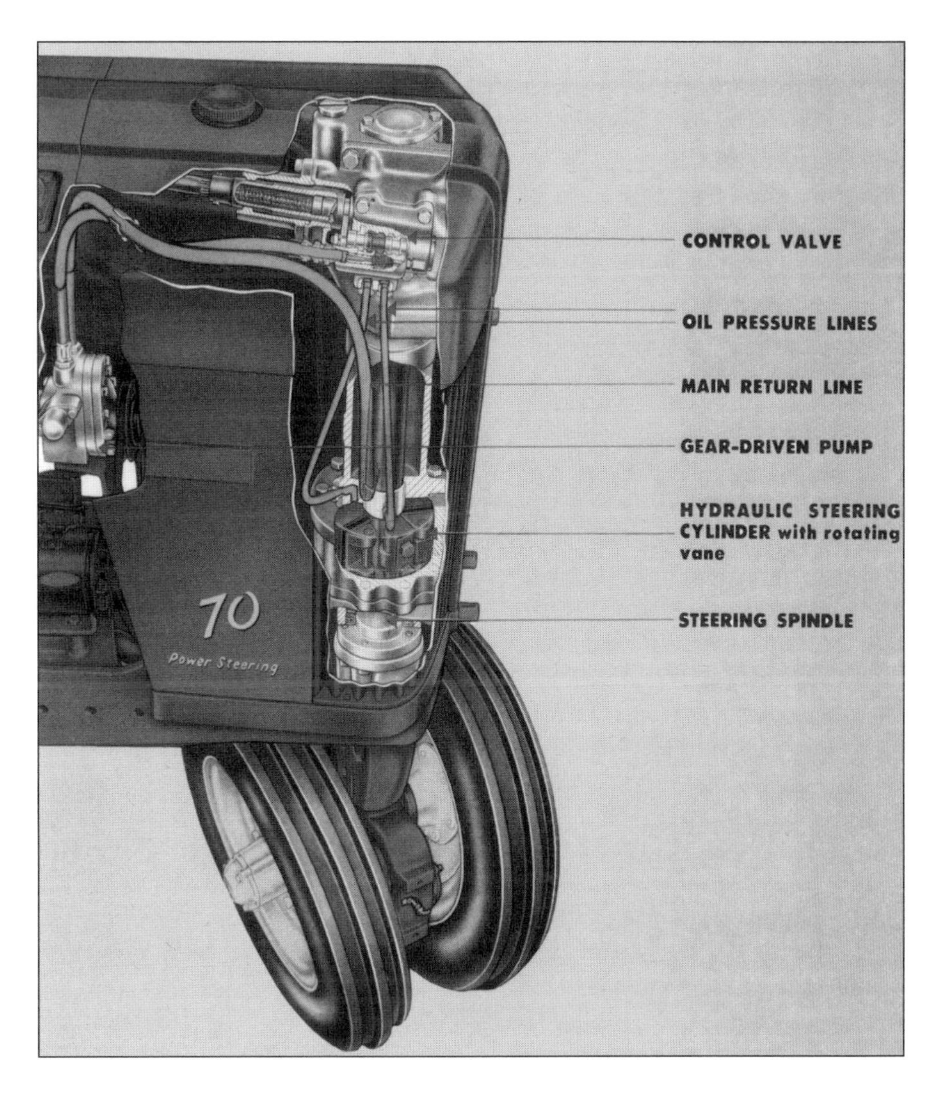

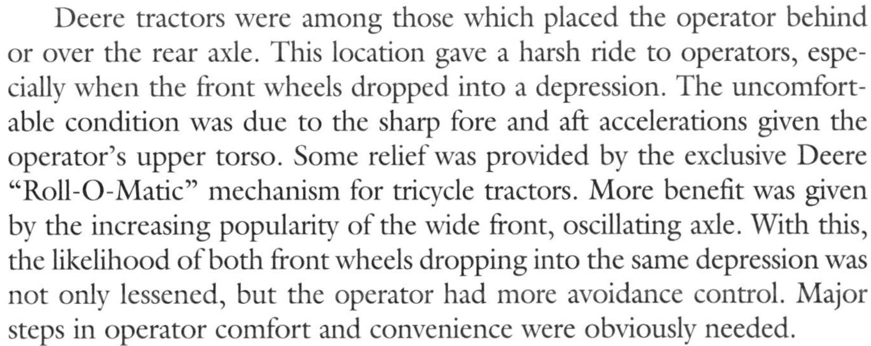

Deere tractors were among those which placed the operator behind or over the rear axle. This location gave a harsh ride to operators, especially when the front wheels dropped into a depression. The uncomfortable condition was due to the sharp fore and aft accelerations given the operator's upper torso. Some relief was provided by the exclusive Deere "Roll-O-Matic" mechanism for tricycle tractors. More benefit was given by the increasing popularity of the wide front, oscillating axle. With this, the likelihood of both front wheels dropping into the same depression was not only lessened, but the operator had more avoidance control. Major steps in operator comfort and convenience were obviously needed.

Another small cloud for Deere & Company was that the same styling appearance and tractor arrangement had been used for many years. Henry Dreyfuss and Associates usually had made some tractor appearance change when there was a periodical redesign. A general feeling existed that a major appearance change would enhance customer attention.

With the above factors (and probably others) being weighed, the company decided in 1953 to begin designing new tractors having four- and six-cylinder engines. This decision was made approximately seven years before production. Deere & Company management were active in supporting and even participating in the design evolution. John Deere tractor history after 1960 attests to the fact that this was the proper action, was timely, and was greeted with considerable success.

The Roll-O-Matic front axle was used on tricycle-style tractors and improved the ride somewhat for the operator.

Decisive Timing by Deere

During the post-World War II period, there was increasing tractor usage and power, increasing field travel speeds, and wider field equipment. These external conditions, combined with internal conditions at Deere & Company, helped prompt the need for decisive action. Most companies, if possible, choose under similar circumstances to seek some way of extending the given current product. Seldom is the bold, entirely new, product choice selected. The reasons Deere & Company made this choice are important to understand.

> *"Some idea of the extent of the human effort required to put the new line of tractors into production can be visualized from the fact that the Deere engineers made 110,000 drawings. This is more than a top-notch draftsman, working eight hours a day, seven days a week, could complete in a lifetime if he was required to make them all himself. The drawings, if stacked one on top of the other, would make a pile more than 44 feet high!"*
>
> — *William F. H. Purcell. ("Industrial Design A Vital Ingredient,"* Automotive Industries*).*

Increased Tractor Use

Farm tractors started to become an important factor in agricultural production about 1920. By that time, the configuration had narrowed to a few principal types. Reliable machines were being marketed and their values in operating cost and convenience were being perceived by the more progressive farmers. Horses declined in number and tractors continued to increase in number until well after World War II.

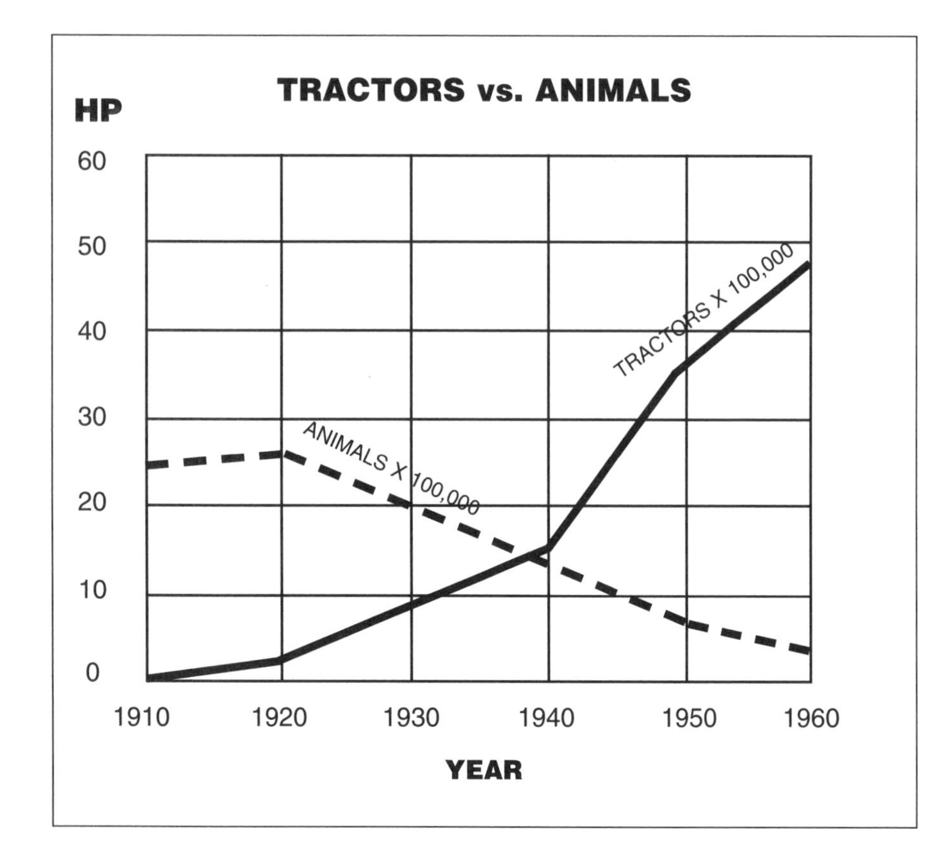

Tractors, not animals, were increasingly used for farm power.

One way to increase tractor power was to connect two units in tandem with hinge steering controls. The operator could control both tractors from the rear vehicle. Note the clock showing the elapsed time for this after-market installation.

Increased Tractor Power

As the number of tractor units increased, so did their power ranges. The top and average power sizes were also increasing. This increased power size was partly the result of declining farm numbers and increasing farm sizes.

Deere managers were aware of customer needs through their service and marketing organizations. Product Engineering personnel often increased their awareness by traveling with service or marketing representatives. This joint travel was to cooperate on field service problems or in random contact with customers to discuss their equipment needs.

Customer contacts were often enlightening. Some customers developed ways to get more out of their tractors. One way was to repower the tractors with engines of increased power over those shipped from the factory. One of the early concepts was the tandem connection of two tractors with a hinge between and steering and other controls provided. In another practice, the factory engine was either customer equipped with higher compression, or larger bore pistons, or both (and later – turbochargers) to increase their power. In some instances the factory engine was completely replaced with a larger power engine, likely from another manufacturer. Some after-market suppliers perceived the desire for more power by some farmers and supplied the adapting parts.

These engine conversions compromised the fuel capacity, cooling system, and transmission durability. They were satisfactory in applications where long use at the new higher power level did not occur. When used for short intervals at higher power, or for reserve power needs, the practice had some acceptance and popularity. These were practices an original equipment manufacturer (OEM) could not use to meet all customer needs. The OEM instead gradually increased the power level of the total machine so all functions were balanced at the given power output.

Increased Travel Speeds

Deere engineers also observed that tractor field travel speeds were gradually increasing. When tractors were mounted on steel wheels, the travel speed (both field and road) was limited by the physical discomfort the operator received. The advent of rubber tires in the 1930s not only gave more operator comfort, but permitted higher travel speeds. The higher field speeds enabled more power to be applied to the implement. Examining this trend was the subject of one Deere research project described as follows.

Deere obtained conversion parts made by an after-market supplier that enabled a 75-horsepower General Motors 2-53 diesel engine to be installed in a 58-PTO horsepower Case LA tractor. With some further modifications, a General Motors 3-53 diesel engine was installed in a Case LA tractor owned by Deere. This engine provided approximately 100 PTO horsepower (approximately the power of the 5010 tractor introduced to the market nearly 10 years later). This tractor was tested by several large farm operators in various geographical areas. Drawbar pull,

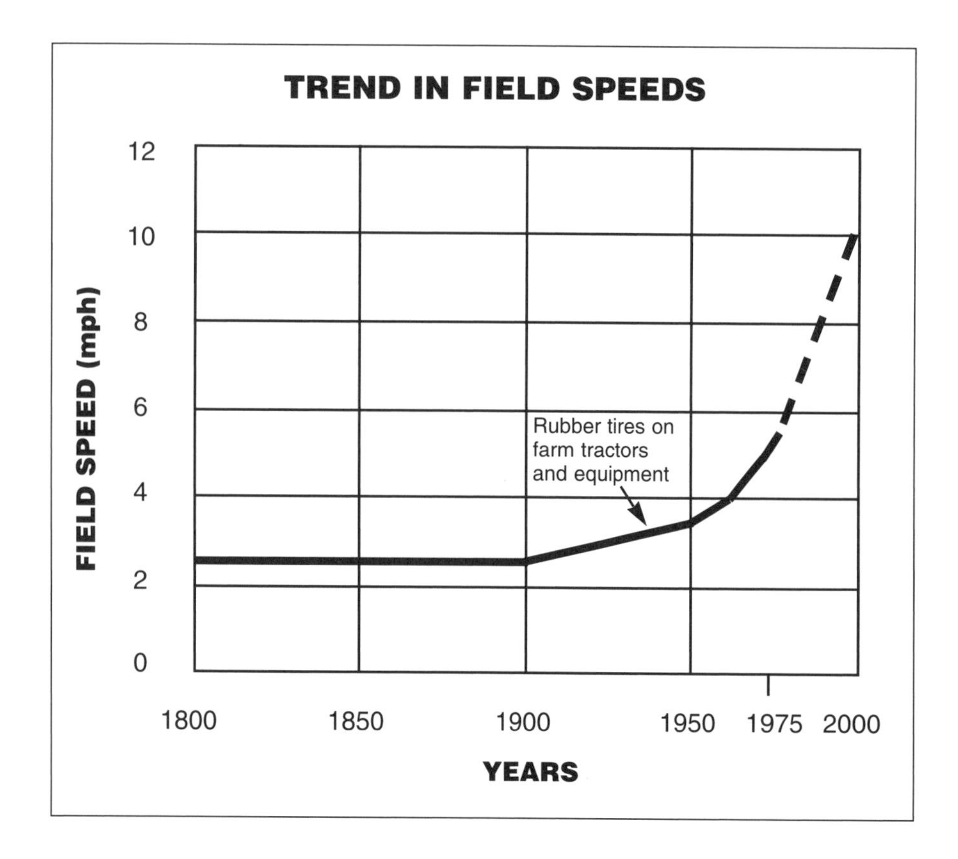

Rubber tires and higher engine power enabled higher field working speeds. The higher field travel speeds permitted more work to be done (Farm Tractors 1950-1975).

travel speed, wheel slip, implement and soil conditions were recorded by field engineers with each test. The operators were asked to use the tractor with their existing or larger drawbar-drawn field equipment in the manner they judged appropriate. Higher field travel was evaluated. The test results supported the belief that higher power could be satisfactorily used. Higher field travel was acceptable, thus requiring the higher engine power.

Powered Operator Controls

Field productivity was increased in other ways besides increasing tractor power. To meet customer demands, the manufacturers began providing wider field equipment. In time, higher field travel limits were reached partly because of increased demands for operator awareness and control, and soil and terrain conditions (for example, fields with rocks). However, field equipment has continued to become larger, prompted by increasing farm sizes and the need for completing field operations in the available time "window."

Increasing field equipment size (both drawn by drawbar and tractor mounted) required the operator to have power assistance to lift the implement. A mechanical lift of mounted equipment first appeared on the John Deere GP tractor in the 1930s. This development was soon followed by the hydraulic lift of mounted equipment (i. e. rockshafts) and hydraulic remote cylinders.

In the late 1940s and early 1950s the Behlen Company in Nebraska made conversion kits that gave power-assisted steering. The first OEM tractors with power steering were introduced by Deere & Company in 1954. This design by Chris Hess, a Deere design engineer, used a vane-type hydraulic cylinder to rotate the tractor's vertical front steering shaft. When the first experimental parts were shown in the engineering shop, Emil Jirsa (then the Design Division manager) predicted its future popularity. This prediction proved true in the coming years when its popularity made it practical to offer the previous power steering option as standard equipment.

This John Deere Model GP tractor was a first use of mechanical power to lift the cultivator rigs.

Factory Equipment Problems

The demands for increased power and power-operator assistance in tractors were arising just as the condition of the factory manufacturing equipment was becoming a concern. Some of the principal production machines had been used for nearly 20 years (beginning in 1933 with the Model A). The per-day factory output of these lower power tractors was greater by as much as double that of the multi-cylinder models a decade or two later. This caused higher relative wear on the factory equipment. Because of their age, the production machines were also less accurate and required more frequent down time for adjustments and repairs.

Simultaneous failure like Oliver Wendell Holmes' "One-Hoss Shay" does not happen. The machines, as they wear, become more difficult to maintain in adjustment, and more discrepant parts are made. If this were only on one machine, corrective steps beyond normal maintenance are possible. However, if many machines are involved, the repair or replacement cost approaches that of re-tooling the whole factory. When this status approaches, it is prudent that the new equipment can manufacture a new advanced product. This allows another long period of customer acceptance.

Power Limit of the Two-Cylinder Engine

Although continuing the two-cylinder engine design was attractive to supporting the existing customer image, several design studies showed the problems of increasing the engine power in the two-cylinder tractors. Tractors, especially row-crop tractors, have a requirement for being able to operate within the majority of crop row spacings. For corn this was principally 38 to 40 inches. Many other crops also used this row width, or a division or multiple of these dimensions.

Increasing the power of piston engines can usually be accomplished by several methods. These methods include improving the efficiency, increasing the displacement (increasing either or both the bore and stroke), increasing the rated operating speed, increasing the compression ratio (spark ignition engines), and adding cylinders. Efficiency improvements are an on-going evolutionary procedure, which alone cannot provide predictable power increases in time or amount. Efficiency improvements are continuously sought by all engine manufacturers.

Increasing the displacement or adding cylinders on the horizontal two-cylinder engines meant the width of the tractor would become larger across the cylinders. This was unacceptable on row-crop tractors, that beginning in the 1930s became much more popular than the standard tread (or wheatland style) models. Increasing engine speed was inherently difficult on the two-cylinder engines because of vibration.

With the two-cylinder engine limited by width restrictions, other options became possible for engine arrangement. A new tractor design would require new factory equipment, but allowed for considerable latitude in concept. The selection needed to achieve customer acceptance, yet still provide long-term growth potential. The Model B tractor, introduced in 1935 at approximately 14 horsepower, exhibited such potential. In 1956 the successor Model 520, manufactured on the same principal factory equipment, had grown to nearly 33 horsepower.

Also, providing a "copycat" design like those available from competitors would give few unique characteristics for a marketing advantage. Mere duplication might not include the latest technology. New and improved designs would give long-term market acceptance in return for the time and money spent.

Decisive Action By Deere

The previously stated factors resulted in a Deere & Company management decision to design and tool a new tractor line. Certain objectives were established, and time was of great importance. On completely new projects, as this was, unforeseen circumstances often cause delays. When the original introduction date of 1958 became doubtful, the '30 series tractor changes were introduced to the two-cylinder models, and the target date became 1960.

The revised introduction date of 1960 was still very timely for Deere & Company. These new tractors were powered to meet the customer's needs at the time and were able to grow with the customer's needs. The customer was entering into an era of unprecedented demand for increasing his tractor's usefulness and productivity. Later features helping their continuing market acceptance were improved transmissions, cabs, powered front axles, and large four-wheel-drive tractors. All of the features were possible with the flexibility of the new designs.

Not only did the new designs meet the prescribed objectives, but in so doing innovative new concepts were designed and used. Some of these concepts were later incorporated in competitors' products.

Ed Fletcher, Project Engineer for hydraulic design, related incidents in those early days: "The New Generation started in a small butcher shop on Falls Avenue. The first thing that Deere did was put paper over the windows. The first thing that the neighbors did was to circulate a petition to prevent the opening of a bar. One not-so-funny event occurred when Morris Fraher, a senior officer with Deere and Company, visited and was standing near the front entry. Naturally, he had security of the project closely in mind. There were procedural changes made after Mr. Fraher opened the door to allow in the morning doughnut delivery man."

Beginning the Design

The initial design work began in April 1953 by setting aside some key persons in a New Design Group. These persons were initially housed in a rented former grocery store (informally called "meat market"). Their work was understood by others in the Engineering Department to be on a future new design. Security provisions (to be later described) withheld details on the specific design or objectives. These security provisions were established with the first small group of approximately 20 persons and were continued successfully as the group became larger.

The project was managed by the late Merlin Hansen, who at that time had about 20 years with Deere, principally in testing and development. His experience and ability greatly influenced the success of the project. He represented the superior product design philosophy stated by John Deere: "I shall never make anything that doesn't represent the best that I can do." Hansen had personal initiative, dedicated commitment, and industry recognition that provided leadership to all who worked with him.

The project leaders had typically come to Deere & Company shortly before or after World War II. Many individuals had the maturing experience that military experience engendered. Most leaders had the experience of growing up on a farm and being acquainted with farm power requirements. They were:

Wallace DuShane, *overall design*; Dan Gleeson, *chassis, seating, controls;* John Townsend, *diesel engine*; John Sandoval, *gasoline engines*; Vernon Rugen, *transmission, final drive, and power take-*

These people developed the new tractors. From left to right, foreground: Harold L. Brock, director of research; Merlin Hansen, chief engineer, new products. Second row: D.R. Wielage, A.W. Lammert, W.H. DuShane, J.W. Seiple, K.B. Sorensen, E.F. Jirsa, B.G. Valentine, W.H. Nordenson, K.J. Harris, W.M. Davis, J.P. Sandoval, and V.L. Rugen.

off shaft; Edward Fletcher, *hydraulic system, including the pump, steering, brakes, valves*; Christian Hess, *implement hitch*; Donald Wielage, *experimental builds and miscellaneous testing*; and Wilbur Davis, *hydraulic system and field testing.*

As time passed and the project grew, many additional employees worked on the project.

The New Product Engineering Center

Land for a new engineering design and test facility was purchased southwest of Waterloo. The first building on the site was made of steel and was informally known as the "tin shed." Occupied in 1955, it housed the first office and shop area. Inside was a machine shop, a welding shop, parts storage, assembly floor, engine test, and other types of support. This building also housed the fabricating, the assembly, and the testing of the first experimental models. The building was erected quickly, without insulation. When winter approached, it was obvious that some supplemental heat was required. A crop dryer was borrowed from the farm headquarters. Without any interior modern plumbing, the crew was obliged to use the outside toilet.

Before the Product Engineering Center farm was purchased a commercial pipeline had been placed across part of the property. Wendell Van Syoc related one incident that this caused a security concern: "The general foreman of the laboratories was very dedicated to maintaining security as well as being obstinate in other ways.

"Therefore, when a pipeline inspector arrived he had problems making a routine inspection. He had great difficulty in convincing the foreman that they had an easement, and that they were going to complete their inspection!"

In the summer and fall of 1955 the first permanent office/laboratory was built. By winter the transmission test wing had been completed. For expediency it was necessary to create a temporary office in the transmission test wing. This enabled test engineers to be near their laboratory — the "tin shed." The evaluation and field test engineers, plus a secretary (Tom Fenelon), also worked in this temporary office (still without modern toilet facilities).

The location and terrain of the farm and the placement of the buildings atop a knoll contributed to the maintenance of security.

One typical Iowa January day, the temperature was below zero and was accompanied by a strong wind. The office became uninhabitable even with extra clothes, so most of the office force was dismissed. They could either go home or complete their day's business at the downtown grocery store office.

This first steel building has since been removed to make space for a shop expansion. The approximately 800 acres have been added to, and the original buildings have over the years been enlarged. By 1965 all product engineering personnel were housed at the site.

This aerial photo of the Product Engineering Center was taken in the 1960s.

General Design Objectives

From the outset it was understood that the new models would replace current models. Several years were needed to provide the new designs, so it was imperative that 1) the current models be updated frequently to maintain their customer appeal, and 2) that information on the new design being prepared should not get to the dealers or customers. This last directive was important to maintain the essential income from current model sales to finance the new venture, which explains the need for good security and secrecy.

"The late president's (Charles Wiman) remark (on April 20, 1953) that it might be only necessary to carry over the yellow and green paint from the old design to the new, was taken quite literally!

"Such courage deserves success." — William F.H. Purcell (Purcell, 1961).

Getting Started

Even before centerlines are drawn on the layouts, company officers, marketing, service, manufacturing, and engineering managers must first reach agreement upon the objectives of the program. These objectives are usually organized by separating the overall, functional, and cost objectives. Examples of overall objectives are the power levels and styling. Functional objectives would be those performance characteristics listed separately for components such as the engine, transmission, and operator station. Various subsidiary objectives and suggestions arose as the design of the tractor proceeded.

Cost Objectives

Most companies use a cost accounting system that compiles the costs of current products and calculates the costs of new product changes. This data would include the cost of the product itself, the investment in manufacturing tooling costs, and product design and development costs. These costs must be assessed before committing a proposed new product to production. An acceptable return on the money invested is necessary for any project to remain viable.

It is common to investigate many alternatives to achieve the desired financial return on the money invested. Deere Waterloo has always had competent cost analysis procedures. A large part of the design effort on the New Generation program was focused on meeting the stringent cost objectives. (The 3010 cost was to compare with the cost of the lower power 70-720-730 models.)

The Deere & Company main office in 1919

Power Level Objectives

One of the first assignments of the New Products Group was to determine an appropriate power level. The overriding performance objective was to give the farmer more tractor for the same or less money, and to ensure that it would do more work with less effort by the operator. The current models being replaced provided some knowledge of the power levels needed for the new designs. A company having only one product in a given market would usually attempt to supply a need in the most popular size. In agricultural tractors, this would be the industry's highest volume power size. A manufacturer fielding several models would want those models to cover a certain power range. It would be important that each model be appropriately spaced in power from the other models. A power spacing too small would mean that each close model would take some sales from another. A large power spacing could mean a power gap between models that a competitor might fill. This scenario is especially true if the competitor sizes his power ratings to fall between the sizes of the first company. In tractors, since power is usually related to price, a large power spacing increases the opportunity for the customer to negotiate his purchase price.

The horsepower of most modern tractors in the United States is quoted at the power take-off (PTO) because it is the most easily measured power outlet. Tractors without a PTO, such as some full four-wheel-drive tractors, use a less easily measured flywheel or engine power. Always, a power rating must recognize some standard temperature and

The Deere & Company main office beginning in 1964.

barometric pressure operating conditions, and these conditions must be stated (see Standards: ASAE S209, SAE J708).

The initial power objectives for the two new models were 50 and 70 PTO horsepower. Engine power included compensation for losses in the drive train from the engine flywheel to the PTO. Power levels have a pattern of increasing during long experimental programs. This program was no exception. When the tractors were tested at Nebraska in 1960 the respective maximum PTO horsepower levels were 59.44 (3010 diesel) and 84.00 (4010 diesel).

Other Objectives

With the power level established, all other important objectives can be derived. From the outset it was recognized there would later be a complete line of tractors, with diesel, gasoline, and LP engines. Creating as many common parts in the complete line was a goal.

Styling Objectives

A meeting with the Moline management on April 20, 1953, suggested these appearance (styling) objectives (Purcell, 1961):

a. The styling should not dictate the design of the tractor. The styling should be functional.
b. The tractor should be green with yellow wheels (a long-time precedent).
c. It should be a "clean design" — not having components projecting beyond the basic sheet metal.
d. The rear of the tractor should be simplified in appearance.
e. A better location (than over the engine as on the then-current John Deere and competitive tractors) for the LP gas (or fuel) tank.
f. The tractor should be more compact for its power.
g. The underside of the tractor should be uncluttered, with the frame along the side of the engine.

Henry Dreyfuss, renowned industrial designer, was instrumental in the styling and design of John Deere equipment for more than three decades.

The Henry Dreyfuss Associates consulted on the styling and appearance. They began this role on Waterloo-built tractors in the late 1930s and continue to this day.

William Purcell, of Henry Dreyfuss Associates, related his first-hand observations (Purcell, 1961) of that April day in Moline:

"What must have been one of the first basic design discussions of this program was held in the rather austere Board Room at the Deere & Company offices in Moline. About 25 people were present including the President, the Vice-Presidents of Marketing, Product Development, and Tractor Production, Chief Engineers of the tractor plants, ourselves (Henry Dreyfuss Associates), and a number of project engineers and product research men. It was an impressive group to say the least, and in retrospect, some of the decisions which were made proved to be pretty sound.

"At this meeting Henry Dreyfuss put in an urgent plea that a place be found for mounting the gas tank, so that when the larger LP tanks were substituted they would not protrude through the sheet-metal hood. He pointed out that the LP tank problem on the present production models, our own as well as those of competitors, could never be solved by an industrial designer; it required an obstetrician! It was decided that we should try placing the tank in front of the radiator.

"There was also considerable discussion at this meeting as to how radical a design change should be made from the present line. After the very radical gas tank decision had been made, the President, the late Charles Wiman, dryly remarked that perhaps the only continuity of design needed was green and yellow paint! — probably the most potent suggestion of the meeting."

Seating and Controls Objectives

Improvements in seating and controls were needed to complement the changes in the transmission and clutch. Superior overall function and comfort were clear objectives. These subjects had extensive investigation with consultant groups arranged by the Henry Dreyfuss Associates (see page 85 for more details).

Transmission Objectives

The transmission had these suggested objectives *(Product Engineering Seminar 1959)*:

- to reduce space occupied by the shifting mechanism, yet the shifting effort should be easy and natural.
- to provide one gear lever movement for shifting either forward to reverse, or reverse to forward, or up and down shifting with a rolling load.
- to consider the use of a foot accelerator for safety in road operation and convenience in field operation.
- to consider safer, more durable brakes.
- to consider a PTO shaft access for the front of the tractor.
- to provide a wider range of speeds than older John Deere tractors.

Suggested ranges were:

	Gear	Travel (mph)
Low speed (creeper)	1	1.5
Convenient speed	2	2.5
Working speeds from	3	3.25
3 1/4 to 5 1/4 mph	4	3.8
with ratio of 1.17 times	5	4.45
next lower gear	6	5.25
High working speed	7	6.5-7
Synchronizing speed (for accelerating highway loads)	8	10-12
High speed	9	15-20

As with all objectives, during the program execution, additional goals were discovered. One such goal was the need for less physical effort and more direct shifting by the operator in the design of the shift levers and quadrant. (See page 39 for more information on transmission design.)

Steering Committee

The product engineers received guidance and approval from a group known as the Tractor Steering Committee. Despite all the humor about products "designed by a committee," no other procedure offers the benefits represented in involving various managerial disciplines.

From the outset it was understood that this would be a very extensive and expensive program and that it would affect the whole company in many ways. The Steering Committee was established to give management control and to communicate with other company management. Membership on this committee included upper management persons with access to the company's various skills in marketing, finance, administrative, and other staff offices. The information transfer process to upper management relied on field demonstrations and periodic organized reviews.

Members of the 1959 Steering Committee were:

Company Vice Presidents

Morris Fraher — Tractor Production
Curtis Oheim — Products Development

Factory Managers

Harley Waldon — Waterloo
Lloyd Bundy — Dubuque

Product Engineering Managers

Merlin Hansen — Chief Engineer, Product Engineering Center
Barrett Rich — Chief Engineer, Waterloo Product Engineering
Wayne Worthington — Director of Research, Product Engineering Center
Floyd Selensky — Chief Engineer, Dubuque Product Engineering

William A. Hewitt, president, Harley Waldon, Waterloo Tractor Works manager, Maurice Fraher, vice president, and Merlin Hansen, chief engineer, admire the new tractor line.

Assistant Chief Engineers
Willard Nordensen — Product Engineering Center
Emil Jirsa — Product Engineering Center
Gordon Valentine — Product Engineering Center
Knud Sorenson — Product Engineering Center

Later, during the experimental tractor period, the name of the group was changed to Policy Committee. This committee continued for some time past the New Generation introduction, gradually evolving into other appropriate procedures and personnel.

Overall Tractor Arrangement

The production model numbers of 4010 and the smaller 3010 were assigned by the Deere & Company Name and Numbering Committee shortly before actual production began. Initially the models were known experimentally as the OX and the smaller OY. (A later program derived the Model 5010 tractor from the OZ experimental designation. The Model 2010 tractor was experimentally designated as the 77.)

The OX was the primary design target. The OX tractor compact design was comparable to the dimensions on the 70-720-730 tractor series. The OY design work borrowed from that already done on the OX and was limited to the unique OY features.

Achieving the OX tractor's compact design included:

- A planetary final drive that saved 10 1/2 inches in length.
- Positioning the "range shift" (high-low-reverse) transmission gearing parallel and in the same compartment and length as the

The layman reader may have difficulty realizing that the space available on a tractor is somewhat limited and subject to compromise. In very early years of the tractor industry some tractor manufacturers purchased many of their components from vendors. Assembled end-to-end, these components determined the resulting tractor length. If the wheelbase was too large, then the front wheel steering was repositioned around other components. Often the operator seat was placed where visibility over the front was poor.

The required radiator size usually determined the operator visibility over the front of the tractor. Behind the radiator, and within the envelope of operator visibility around the radiator, there was usually space behind and above the engine, making this a common location for the fuel tank.

The early tractor design established the wheelbase, the design volume, and the factory tooling fixtures. Changing the overall design, including the wheelbase, to obtain more design volume often meant changing many parts and tooling, affecting both the tractor and the factory. Designers were reluctant to change the vehicle size with an increase in power. As time passed, the fuel tank was not large enough because the usual space behind the engine did not increase with the increased engine power. Experienced tractor designers know an acceptable tank size to provide adequate field running time before the need to refuel.

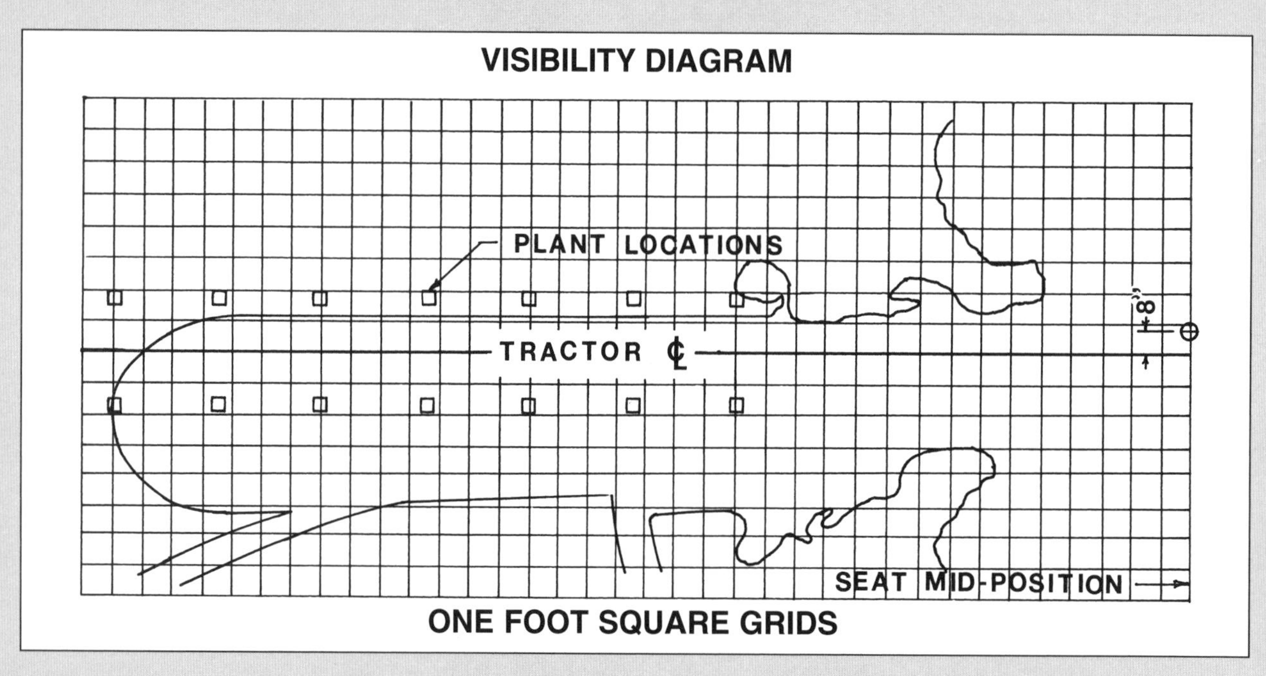

Plot of shadow projection of tractor on grid in darkened room. Single point light source simulates operator eye at 8 inches to right of tractor centerline.

"speed change" gearing. Compared to the conventional design with one compartment ahead of the other, this configuration saved 10 inches in tractor length.

These concepts saved a total of 20 1/2 inches in length. Other advantages included:

- allowing the engine to be closer to the rear wheels,
- reducing the prospect of overloading the front wheels, and
- providing ample space for the front-mounted fuel tank. (The short coupled engine-transmission concept was useful in the OY crawler tractor version.)

The OX wheelbase, weight distribution, and operator visibility were compared with the existing production model. Some of the pertinent Nebraska Test comparisons follow:

	730	4010
Nebraska Test Number	594	761
Maximum Test Weight (lb)	9,241	9,775
%/Weight (lb) front wheels	22.1/2,046	23.3/2,280
Minimum Test Weight (lb)	7,861	7,445
%/Weight (lb) front wheels	25.9/2,040	30.9/2,300
Rear Tires in Tests F, G, H, and J	15.5-38	15.5-38
Test K	13.6-38	
Front Tires	6.00-16	6.00-16
Wheelbase (in.)	90.9	96.5
Max. PTO Horsepower/Engine rpm	56.66/1,125	84/2,200

Front Fuel Tank Placement

A typical first observation was: "Why was the fuel tank placed in the front?" The suggestion by Henry Dreyfuss led to an intensive study to explore and justify relocating the fuel tank.

Fuel tank studies, comparing the front location and the conventional behind-the-engine location, were conducted. These studies showed the front location provided greater volume. The 4010 was a tractor nearly equal the 830 model in power but compacted into the 730 volume. Hence, the new design needed a larger proportioned fuel tank to enable the larger power engine an acceptable time in the field before refueling.

This new tank placement concern was further increased with the liquefied petroleum gas (LPG) fuel models. This fuel, with less energy density, required a cylindrical-shaped tank for strength to withstand the storage pressure. Therefore, these tanks were larger and harder to conceal within the styled sheet-metal profile. A compromise was reached between the amount of tank exposure to visibility and field running time. Running time data follows for these front-located tanks. This showed the

The LP gas fuel tank arrangement of the 630 LP tractor.

This photograph shows the LP gas fuel tank arrangement of the 4010 LP tractor. This projection beyond the hood was needed on the 4010 tractor but not on the 3010 tractor.

effect of staying within (under) the styled sheet metal, or projecting above the hood line (time of operation at full power):

LPG Model	Under Hood Time	Above Hood Time/Amount
OX	3.97 hours	5.18 hours/7 13/16 inches
OY	4.77 hours	5.65 hours/5 1/8 inches

The OY, at 4.77 hours with no above-hood projection, was nearly equivalent to the OX with 7 13/16 inches of projection. This data permitted the 3010 tractor LP tank to not project above the hood.

Secondary advantages of the front fuel tank placement were:

- reduced fire hazard from spillage caused by leaks or refueling,
- reduced gasoline loss by evaporation,

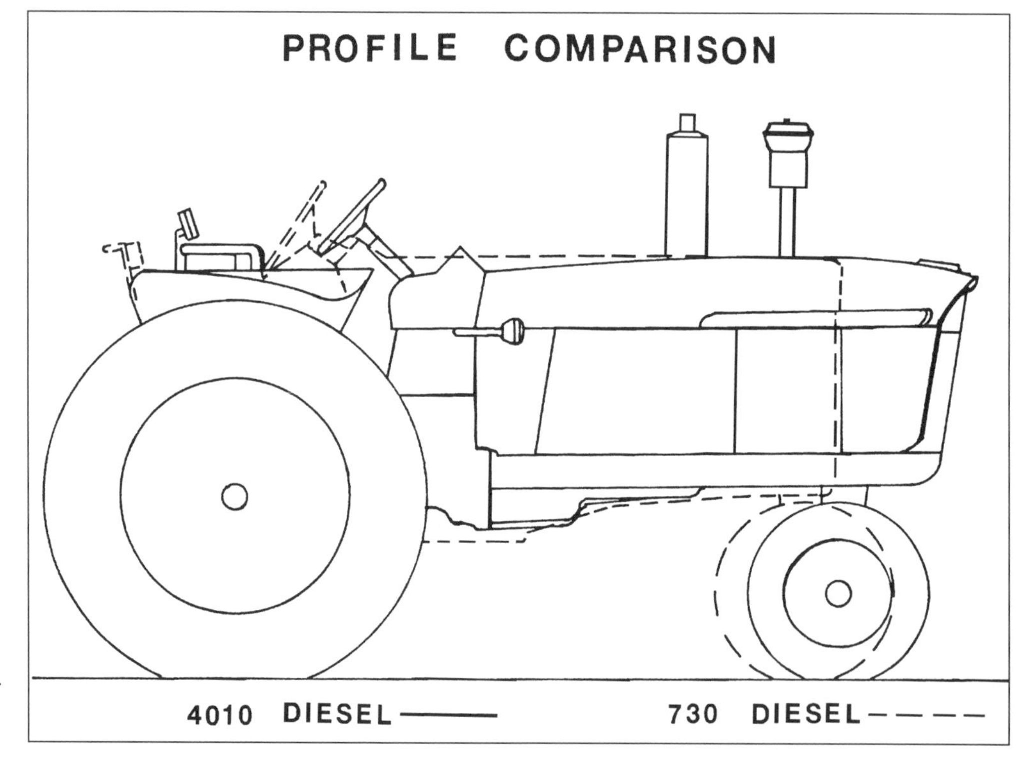

This drawing provides a good comparison of the profiles of the 4010 and 730 tractors.

- increased access for engine servicing,
- eliminated special insulation for the gasoline, diesel, or LPG tanks, and
- enabled a sturdier and less expensive tank mounting.

With all innovations there are both advantages and disadvantages. One disadvantage of a front-mounted fuel tank is the variable weight resulting on the front wheels. However, it can be mathematically shown that the variation is not a large percentage of the total weight and is well within the front axle weight variations from other causes.

Another disadvantage is that placing the engine closer to the operator causes higher heat and noise. This item was the subject of considerable investigation to provide satisfactory operator comfort.

Operator Station

Compared to the 730 tractor operator position from the rear axle, the OX design located the seat 11 1/2 inches forward and 1 1/2 inches downward. This location provided more room for the hitch components. Importantly, it placed the operator closer to the center of oscillation in the longitudinal plane, giving a more comfortable ride.

The platform was flat for its full width. The front portion was turned upward to provide a foot rest. The seat could be easily moved out of the way if the operator wanted to stand.

Much study was given to control placement. It was important that the controls were adjacent to the steering wheel. Also, they needed to be within the operator's sight line when he was seated and steering the tractor.

The tractor could be mounted from either the left or right, or front and back of the drive wheels. Access from the rear was not as convenient since it was crowded by the standard equipment fender and the rockshaft arms. Therefore work was done to make the front entry and exit more attractive with the addition of steps and hand holds. (Rear entry was later given up entirely with the advent of cabs.) Better seat cushioning was provided. It was placed upon a suspension that had improved ride and adjustment for a range of driver sizes and weights.

From the basic objectives the design engineer began producing the concepts. Obviously, some variation in guidelines was found necessary as the design (and testing) progressed, but the main overall, functional, and cost objectives remained as firm guidelines.

The New Generation deluxe posture seat and suspension.

Engines

Farmers often have strong emotions when it comes to their tractors. The performance of the engine is a strong influence on the farmer's opinion. For many operators, the engine is what gives the tractor its personality, its character, its spirit. A satisfied tractor owner may say his tractor has good lugging ability, is robust, or is strong. Those statements imply to an engineer that the engine has a "good" torque curve, meaning that torque increases as engine speed is reduced below rated speed.

Cold starting characteristics were tested in these cold room laboratories.

Not having to shift to a lower gear when a momentary load increase is encountered is important. Likewise, it is important to not have to shift to a higher gear when the momentary load increase is past. Time is not only saved, but principally, the required operator effort is reduced. Low shifting force is an objective of transmission designers, but excessive frequency of shifting is an effort by operators. Hence, the need for an adequate engine torque reserve. The speed interval over which the torque continues to rise should exceed the transmission speed spacing.

Other engine characteristics or qualities help make operators pleased with their tractors. One is the capability of "cold-starting" in low temperatures. Another is the frequency and amount of routine maintenance necessary to maintain good operating condition. Low breakdown frequency is highly important, especially in the busy season (when most failures occur). A farmer is always under a deadline to get field operations completed in the finite window of time that weather and crop conditions allot him. A dependable tractor helps toward overcoming that deadline anxiety.

This artist's illustration shows a tractor cresting a small field incline. Tractor engines were expected to have reserve torque characteristics that permitted them to work modest inclines without shifting the transmission to a lower gear.

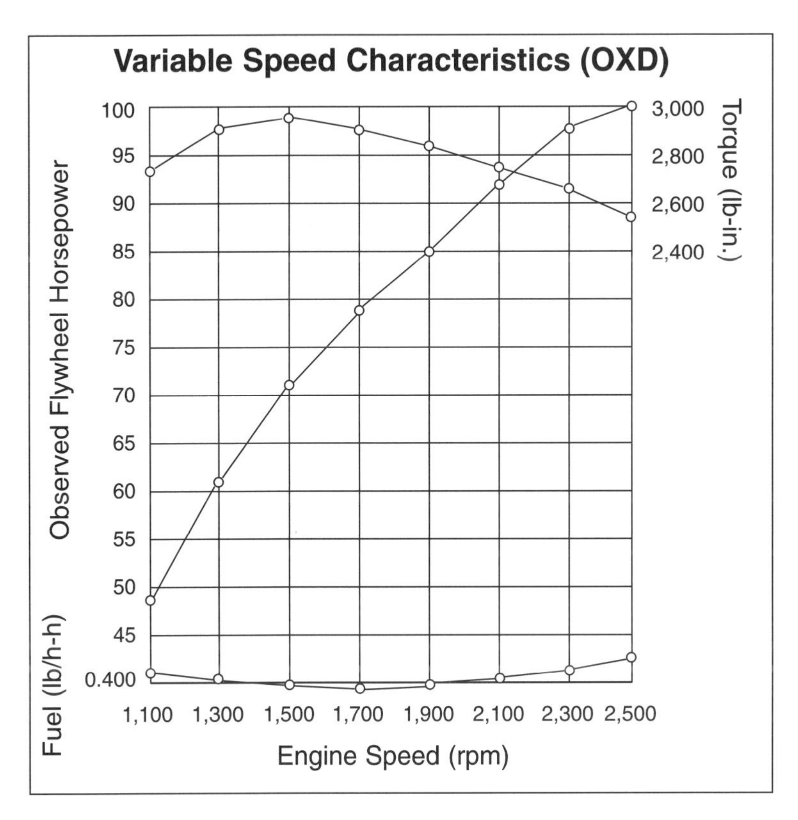

Torque and horsepower curves are the most common graphical illustration of engine performance.

Most operators have mechanical ability and understanding of internal combustion engines. These traits have helped tractors achieve their general acceptance. This mechanical ability has helped many tractors continue to be productive without always awaiting a dealer's call when a minor problem arises.

Low overall cost is also important. A farm operation is a business. Tractors and tractor operation are part of the costs of doing business. Farmers must take into account the cost factors of purchasing and operating tractors and implements. Hence purchase cost, reliability, durability, fuel and maintenance costs are priorities when purchasing a tractor.

Engine and Tractor Design

Accomplishing the aforementioned engine characteristics were some of the goals of the product engineering group. They needed to work as a team and involve the manufacturing, marketing, service, and company management as appropriate to meet their goals.

The following discussion depicts detail engineering activities on engine design and testing. Similar procedures may have occurred for other tractor components but they will not be described with the same detail.

First, the product must have the specifications to meet the intended market. These specifications for engines include power, torque rise, service intervals and cost, noise, fuel consumption, cost, and weight. Once the desired features are specified, the design process begins. Parts were designed to a conservative stress level yet were consistent with the cost objectives. This helped decrease future down time problems in operating the parts, both in experimental testing and customer use.

Experimental parts were made according to the drawings, then were assembled and tested. Some parts such as a connecting rod, crankshaft, or a cylinder block could have exhaustive testing as individual parts. Further testing may be made as assemblies, such as a whole engine. The availability of computers in recent years has reduced the time spent on the actual parts testing.

Tests were generally separated into two classifications: performance and durability. Experience with other production units was used to set the specifications of power, fuel consumption, and durability. Typical performance tests included the power developed by an engine, flow and pressure characteristics of a hydraulic pump, and others. Several types of tests were used to achieve the durability objectives. These may have included components tested at a certain number of hours at given load levels, or a period of testing in complete tractors. Tractors may have been tested in company facilities or in selected customer usage.

Development was known at Deere & Company as the combined task of redesigning and retesting. Throughout the design and testing phases, manufacturing, service, and reliability departments were represented.

Separate testing of all individual parts would have been costly and time consuming. Individual testing might have been done on such complex parts as: connecting rods, hydraulic pumps, seat suspensions, or the shock absorbers used in a seat suspension. The evaluation of design changes could be done more precisely with repeatable and controlled laboratory testing.

Testing complete tractors in actual field tests provided valuable information since it simulated customer usage. However, the variable load in field testing took longer and the failures encountered along the way could not always be repeated.

They not only made recommendations for design improvement, but also proceeded with their responsibilities for production readiness.

In the days of the New Generation testing, a common procedure was to build the parts and then test them. Stop-action (often called strobe) lighting could detect valve bounce or spring vibration. Excessive camshaft wear would show with durability testing by either actual wear or wear pattern, depending upon the amount of running time. Inlet and exhaust pressures and temperatures were recorded for comparison with the design parameters.

Achieving the design objectives at the first testing was unlikely. Hence, many hours of test and retest were expected and used. Multiply this by all the operating specifications an engine must have and it is understandable why developing major engine (or complete tractor) changes are not frequent or done quickly. One objective of an engineering group is to complete the assigned task. Another often overlooked objective is to complete the task more efficiently.

It is not the object in the following descriptions to be exhaustive in covering all aspects of the engine features. Rather, the discussion intends to concentrate on the principal and the unique (at the time) concepts, and the investigative work.

Engine Operating Specifications

The Model 530-830 tractor series all had the rated PTO (power take-off) speeds occurring at the maximum (rated) engine operating speeds. The New Generation tractors had 1,900-rpm engine speed for the rated PTO speed while the engine operating speeds varied from 1,000 to 2,500 rpm. The throttle control had an override detent to mark the 1,900 rpm for use with many PTO-powered equipment.

The variable operating speed enabled the operator to "shift up and throttle back" to gain better fuel economy and less noise on light loads. Although these were higher power tractors, the need for slower engine speeds could arise. This may be the only tractor the farmer had available for light loads such as mowing.

The hand throttle position for 2,200 rpm allowed more power for non-PTO field usage (often the greater usage). A foot speed override control was used in road travel to achieve the 2,500 rpm and higher travel speeds. A combination of the foot accelerator and transmission shifting could be used to get rolling loads up to highway travel speeds.

A comparison of tractor PTO (or belt) powers as developed at the Nebraska Test is shown below:

Model	1,000 rpm PTO	2,200 rpm Engine
3010 (OY)	55.29	59.44
4010 (OX)	76.71	84.00

Model	1,125 rpm Belt Horsepower
630 G	46.75
730 D	56.66
830 D	72.82

Example of Design and Development Refinements

Some readers may be unfamiliar with the methods used by the engineer to achieve various objectives, such as fuel consumption, torque reserve, or any other operating characteristics.As an example, consider achieving a torque reserve goal. Suppose the objective is to increase engine torque from rated speed to 70% of the rated speed. The result is that at the peak torque speed there must be ample fuel and air for proper fuel combustion.

Controlling the fuel amount is the function of the injection pump (or carburetor on spark-ignition engines). The injection pump (or carburetor) is controlled by the engine governor and the operator control lever. Fuel is metered through the fuel injection pump (or carburetor) as demanded by the load, operating speed, and throttle position.

Insufficient air means lower horsepower (torque times speed) and excessive exhaust

Wendell Van Syoc, a field test engineer, related this test experience: "On one occasion at Laredo it was necessary to check the performance of an OX tractor in rough ground. (Carbureted engines can sometimes misfire if the tractor motion disrupts the fuel mixture.) The tractor was driven across a freshly bedded field. The muffler was not clamped tightly to the exhaust pipe. On one severe bounce the muffler flew off and landed in the loose dirt.

smoke. Getting adequate air in or out of the engine cylinder is not as easy as merely increasing the intake or exhaust valve size.

A larger valve could cause problems in other mating parts. The greater valve mass could overload the camshaft, the valve spring, and the rocker arm levers. This situation could require size changes in the whole intake and exhaust system, not to mention finding room in an already crowded cylinder head. Attention would also be given to adjusting the cylinder head valve lift and duration, reducing the air passage restrictions, and other details. A valve opening too fast could cause excessive stresses on the cam profile faces, valve bounce, and valve spring vibration. The design study that derived the best compromise of all the above characteristics must then be built and tested. Today, most of this can be simulated on computers, thus lessening the sequence of design, layout, build, test, revise, and test again.

"One person who was wearing leather gloves picked up the hot muffler and placed it back on the tractor. The engine would not crank. Since it was near the end of the test program the disabled tractor was pushed onto a truck returning to Waterloo. When the cylinder head was removed at the Engineering Center a plug of dirt that had been inhaled from the muffler was found on top of one piston."

The 5010 tractor, introduced in 1962, had different engine piston and connecting rod assemblies from the 3010 and 4010.

Three basic engines were designed with a maximum of interchangeable parts. The first two in production were for the 3010 and 4010, with the 5010 being introduced later in 1962. The 3010 and 4010 shared the same piston and connecting rod and the 5010 was different. Initially, the three models were intended to have gasoline, liquefied propane gas (LPG), and diesel engines. For the 5010, only the diesel engine reached production.

Design History

Extensive prototype engine study had been conducted before the formation of the new tractor design group in the former "meat market" in Waterloo.

Vertical two-cylinder in-line diesel (2) and gasoline (1) test engines were built. This engine series, called OX 21, had a 3-3/4-inch bore and 4-3/8-inch stroke and was rated at 1,800 rpm. The test results from these engines provided much of the information for designing the later four- and six-cylinder engines.

One of the major features tested was the Hartford Machine Screw Roosa Master (now Stanadyne) rotary diesel fuel injection pump with the distributor-type drive. Early in 1953, a decision was made that the first designed-and-built engine should be approximately 45 horsepower. The overall size was to be approximately that of the Model 60 (gasoline — 41.18 horsepower) which it would replace. The diesel, gasoline, and LPG engines would share a maximum number of common parts.

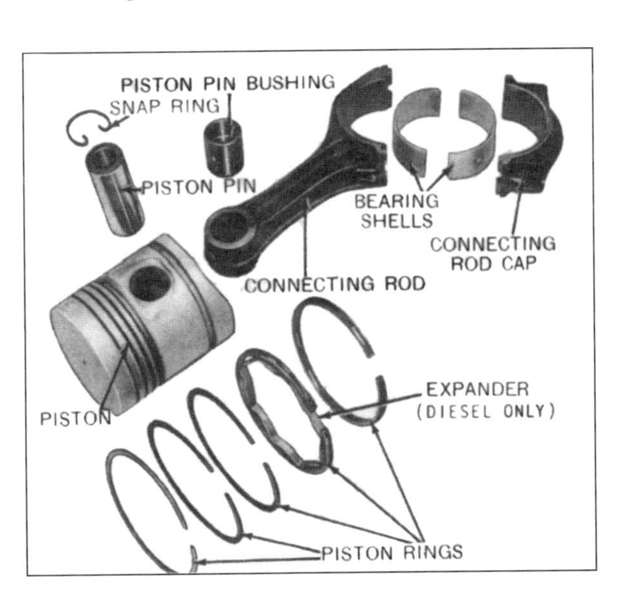

This view shows the disassembled 4010 (and 3010) piston and connecting rod assembly.

The engine groups studied these engine arrangements:

- an in-line four cylinder,
- a 60° V-type four cylinder,
- a 45° V-type four cylinder,
- an in-line four cylinder with an overhead camshaft, which was favored for initial work.

As design layouts began, many other factors (such as market difference) arose. These developments caused the four-cylinder in-line concept to be abandoned for the 45° V-type six-cylinder of approximately 70 horsepower. The power objective then became the Model 80 tractor horsepower (65.33 horsepower).

A 45° V-type six-cylinder design was the first engine built. It had a 3-3/4-inch bore by 4-inch stroke, rated at 2,100 rpm. Only one, a gasoline engine, was built with both the exhaust and inlet manifolds on top of the engine between the V.

The smaller size, lower cost, and more simplistic Roosamaster rotary distributor diesel fuel injection pump is compared to the in-line design in this photograph.

A second V-type design was made in both four-cylinder and six-cylinder versions. These engines had the above bore and stroke but only the intake manifold was within the V, and the exhaust manifolds were outside the cylinder banks. Five four- and six-cylinder engines were built for testing. Manufacturing costs and the increased width caused the cancellation of further V-type engine consideration. Structural and vibration problems caused inferior performance. Work then resumed on in-line engines with a structural design capable of handling future diesel engine power levels.

This early styling rendition of the OY tractor used the V-4 engine. Note the outward projection of the engine cylinder heads.

Evolution of Experimental Engine Design

- *Vertical two-cylinder gasoline and diesel test engines (3-3/4-inch bore, 4-3/4-inch stroke, with 1,800-rpm rated speed).*
- *Design of in-line four-cylinder engine with overhead camshaft (design only).*
- *45° V-type four- and six-cylinder engines (3-3/4-inch bore, 4-inch stroke, with 2,100-rpm rated speed).*
- *In-line four- and six-cylinder engines called the XL series. (Gasoline and LPG bore and stroke was unchanged, but the diesel had a 4-inch bore and 4-inch stroke.)*
- *Refinement of the XL engines with relocated driven accessories (known as the program 5 and program 7 designs).*
- *Redesign of the programs 5 and 7 engines to change the engine rotation. (4-inch bore and 4-inch stroke for the gasoline and LPG engines, 4-1/4-inch bore and 4-inch stroke for the diesel engines. Rated speed changed from 2,100 to 1,900 rpm on all engines.)*
- *Diesel engine bore and stroke changed to 4 1/8 inches and 4 3/4 inches, respectively. These fourth program designs were called program 8C (OX six-cylinder) and program 9C (OY four-cylinder).*

The first designed and built in-line engines were called the "XL" series. The gasoline and LPG engines had a 3-3/4-inch bore by 4-inch stroke. The diesel engines were 4-inch bore by 4-inch stroke.

A second series of engines, experimentally called programs 5 and 7, had the same bore and stroke dimensions. The camshaft, oil pump, generator, governor, distributor and fuel injection pump drive were all relocated. These new locations mostly continued into the production engines.

The next (third) series revised the gasoline and LPG engines to a 4-inch bore by 4-inch stroke, and the diesel to 4-1/4-inch bore by 4-inch stroke. The rated speed of all engines was reduced from 2,100 to 1,900 rpm. This alteration agreed with the philosophy that the rated PTO speed would be at the 1,900-rpm engine speed. With other changes the engine rotation was changed from counter-clockwise, when viewed from the front or fan end, to the conventional clockwise. Counter-clockwise engine rotation had originally been chosen to reduce the number of gear shafts needed in the transmission to obtain proper PTO shaft rotation. The counter-clockwise rotation would have required adapting other types of equipment in the engine — which led to its demise. The primary commercial use of counter-clockwise rotating engines is in twin engine marine applications. This engine rotation enables two opposite rotating propellers to be used to avoid sideways motion of the boat.

The 4-1/4-inch bore by 4-inch stroke was what is commonly termed an "over square" design; that is, the bore diameter is greater than the stroke length. This diesel engine could not be successfully developed to provide the desired performance. It was unsatisfactory in operation and starting. Poor starting occurred though the compression ratio was increased to 19 to 1 to compensate for the increase in cylinder heat loss and lack of compression pressure. This was further proof of the test results found with the OX 21 engines.

The fourth program 8c (OX) and 9c (OY) engines were refined and released to production for the 4010 and 3010 tractors. A new bore of 4 1/8 inches and stroke of 4 3/4 inches was designed for this diesel engine. This change provided a longer compression stroke and better scavenging of the cylinder during the exhaust cycle. The spark ignition engines continued at the previous design of a 4-inch bore and a 4-inch stroke. When the bore and stroke measure the same, the term "square" is sometimes used.

Engine Balancing

On a four-cylinder in-line engine, the crankshaft configuration moves the two end pistons opposite the direction of the inside pistons. The inertia forces of one reciprocating piston and connecting rod are then partly cancelled by another piston and connecting rod. The inertia forces are not completely cancelled because the pistons travel faster during the upper half of the stroke than during the lower half. This difference in acceleration of two opposing piston sets causes unbalanced forces to occur at twice engine speed.

A balancer assembly was provided on the four-cylinder engines to compensate for these unbalanced forces. The balancer was driven by a gear welded to the crankshaft between the number two crankshaft throw and the center main bearing. An intermediate gear was used to reach the balancer assembly placed below the crankshaft in the crankcase. This gear rotated one of the two balancer gears at twice engine speeds.

The engine balancer assembly effectively cancelled the unbalanced vertical forces in the four-cylinder engine.

The balancer gears had offset weights in their webs and were gear driven and timed with each other. The offset weights were oriented such that they cancelled their unbalanced effect in the horizontal plane, but in the vertical plane their effect was additive. This vertical unbalanced force was timed with, and effectively cancelled, the vertical inertia forces previously described occurring at twice engine speed. Students of modern automobile engines will recognize this as the so-called Lanchester vibration damper in use by some makes (Honda, Mitsubishi, etc.).

In the six-cylinder engines, torsional vibrations exist due to the inherently longer crankshaft design. A torsional damper was used in the front crankshaft pulley to isolate these torsional vibrations from the fan, generator, water pump, and hydraulic pump.

Electrical Components

These engines used generators similar to those used on the two-cylinder tractors. It may seem strange today to realize that in the late 1950s there were to be several more years before alternators came into use on agricultural tractors. In fact, they were not yet in universal use on automobiles. Gasoline and LPG engines used 12-volt generators while 24-volt generators were used on the diesel engines because of their need of the higher starting power provided by 24-volt starting systems.

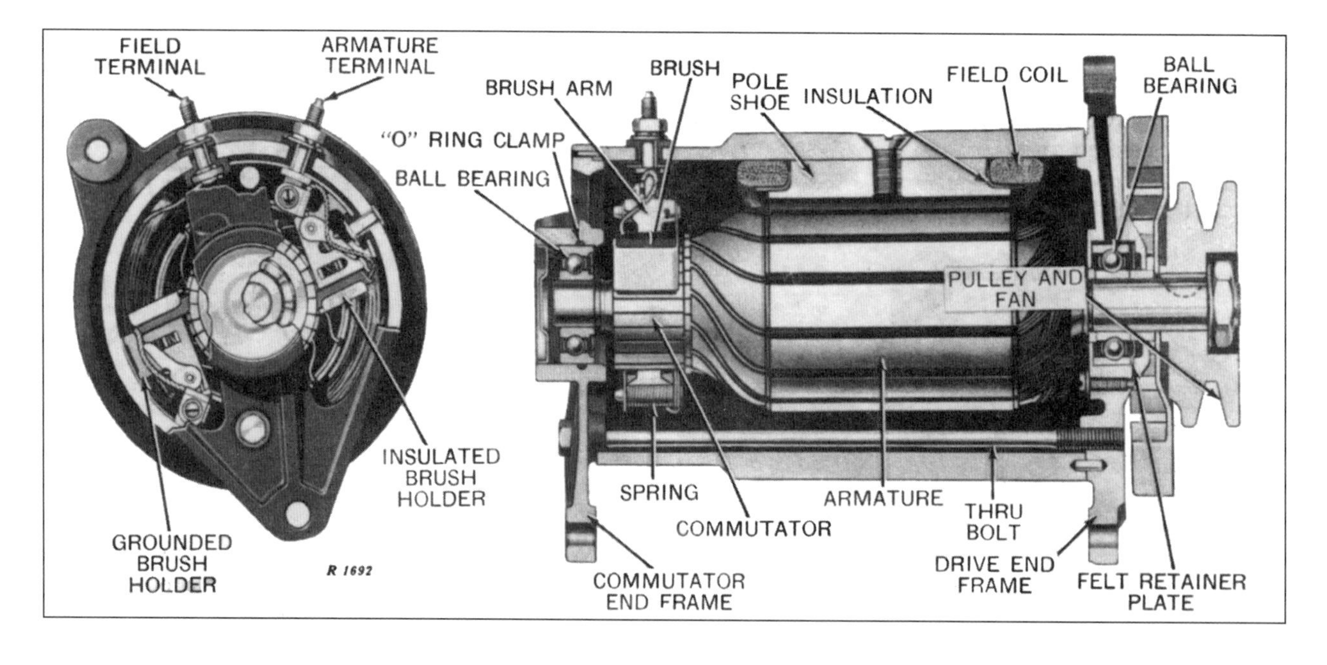

This sectional view shows the inner workings of a 12-volt generator.

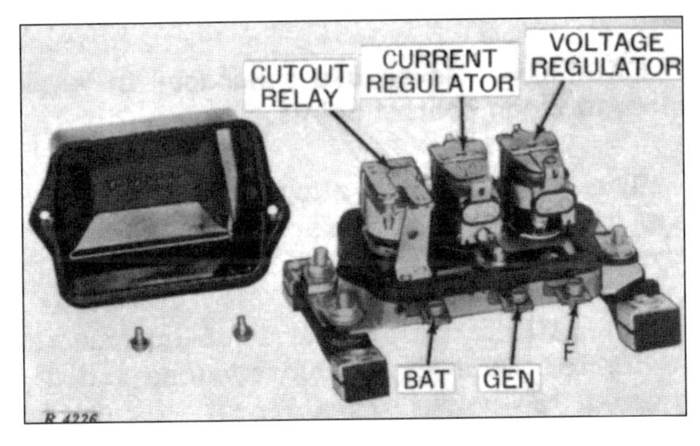

This picture shows the three unit voltage and current generator regulator.

The principal generator change for the new tractors was intended to provide sealed ball bearings on both ends. The rear sleeve bearing, as used on the older tractors, were subject to early failure at the higher speeds of the 3010 and 4010. The following table documents those speeds:

Model	Rated Engine Rpm	Generator Rpm
3010/4010	2,200	4,700
520	1,325	3,260
620/720	1,125	2,950

The 24-volt regulator was the same three-unit version used on the 730 and 830 diesel tractors. The 12-volt regulator was changed from the two-unit version used on earlier tractors to the three-unit design to limit generator output. Had the two-unit regulator been used, output could have exceeded 30 amps, well above the generator rating of 20 amps.

The two-cylinder tractors used a manual engagement of starter pinion and relay contacts, instead of the more convenient electric solenoid. This choice was because of the low location and the proximity to dust, water, and mud. On the 3010 and 4010 tractors with the high forward generator location, the more convenient electric solenoid engagement of the starter pinion and relay contacts was found trouble-free.

As a starting assist, the 3010 and 4010 main hydraulic pump, driven off the front of the crankshaft, had an unloading feature. This feature had awkward manual access and was probably seldom used. However, it did help significantly to attain a higher cranking speed in cold weather conditions.

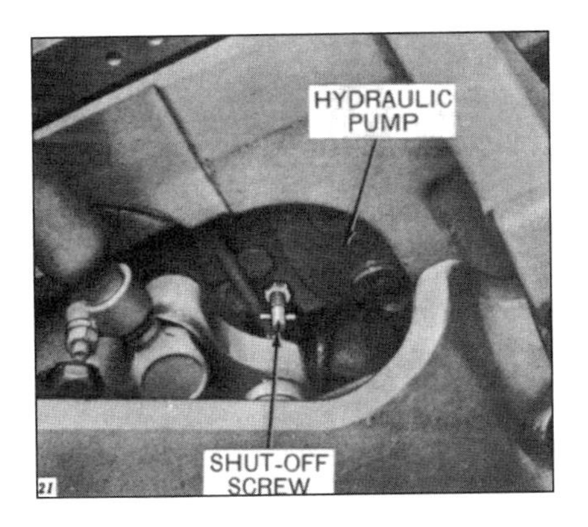

The early design hydraulic pump shut-off screw was accessible through the bottom frame support.

Crankcase Ventilating System

Positive air circulation in the crankcase had been used by the two-cylinder tractors for more than 10 years. The pump for this system was either a vane or roller type, mounted and driven off the front of the fan drive shaft.

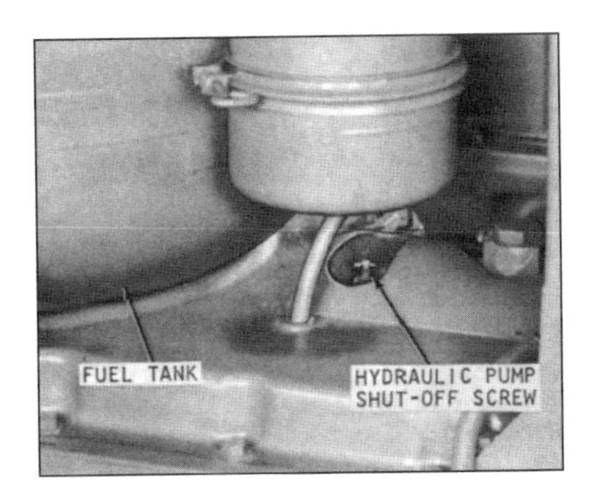

The later design hydraulic pump shut-off screw was more easily accessible through the upper frame plate.

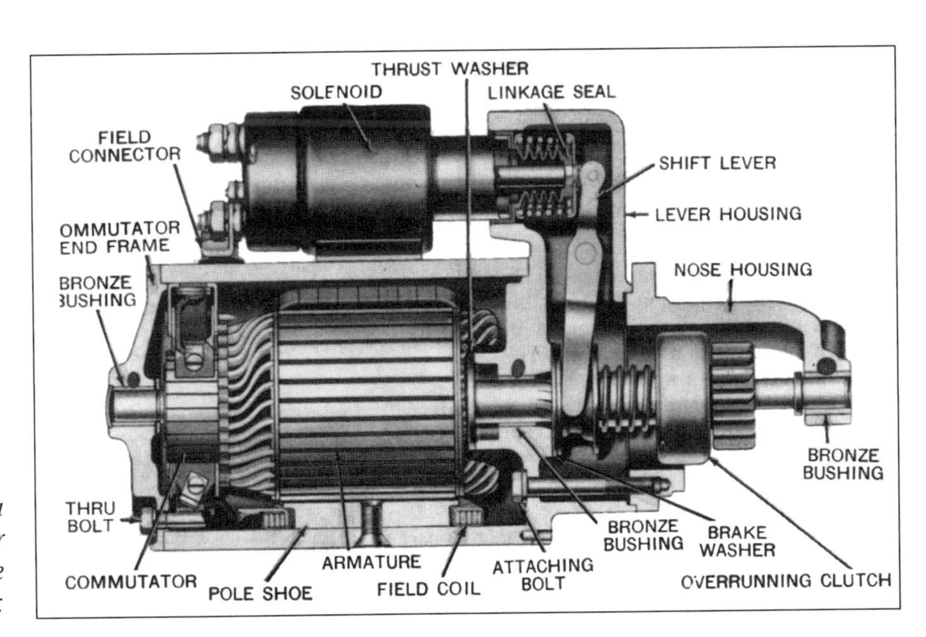

This sectional view of a 24-volt starting motor provides a good look into the inner workings of a motor.

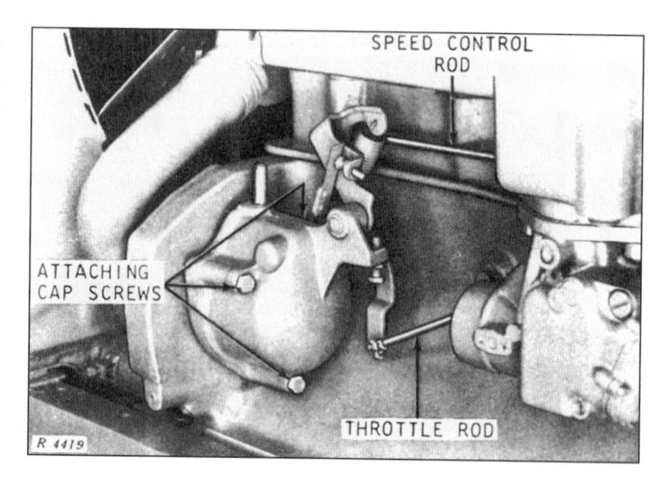

The governor and ventilator pump housing are on the spark ignition engine.

A new, simpler pump was designed for the new four- and six-cylinder engines. This pump had a vaned aluminum rotor revolving in an offset housing chamber that was partly filled with crankcase oil. With rotation, the oil approached and receded from the pump rotor hub. This acted as a liquid piston to displace the air trapped in the rotor. Air was conducted through the pump by appropriately placing the inlet and outlet air passages. Approximately 70 cubic feet per hour of air was taken from the clean air system. The air was discharged into the crankcase at the front of the engine.

The pump created approximately 0.2 inches of water pressure. It was sufficient to force crankcase air and blow by gases up through passages in the cylinder block into the tappet lever cover. From there it passed out a tube, similar to a road tube on trucks and automobiles, attached at the rear of the cover. This tube discharged the flow below the engine frame just ahead of the right side of the flywheel housing. The pump had no failures during more than 100,000 hours of engine operation by the start of production.

Gasoline Engines

Naturally, there were other differences between the spark ignition (gasoline and LPG) and the diesel engines, besides the bore, stroke, and crankshafts. As may be expected, the gasoline tractors used carburetors to measure the fuel delivery to the engine intake manifold. Initial testing with both single and dual downdraft carburetors was unsuccessful due to the higher ambient air above the engine when enclosed with the hood. Therefore, work turned to developing updraft carburetor designs. Both Marvel-Schebler (Borg-Warner) and Zenith (Bendix Aviation) carburetors were released to production. The two carburetors were different in their accelerator pump, idle passage connection, part load economizer, and, of course, the actual castings. Two carburetor suppliers were provided to lessen the dependence on only one supplier.

The Marvel-Schebler gasoline carburetor was made by Borg-Warner.

The Zenith gasoline carburetor, made by Bendix Aviation, was also used to lessen the dependence on just one supplier.

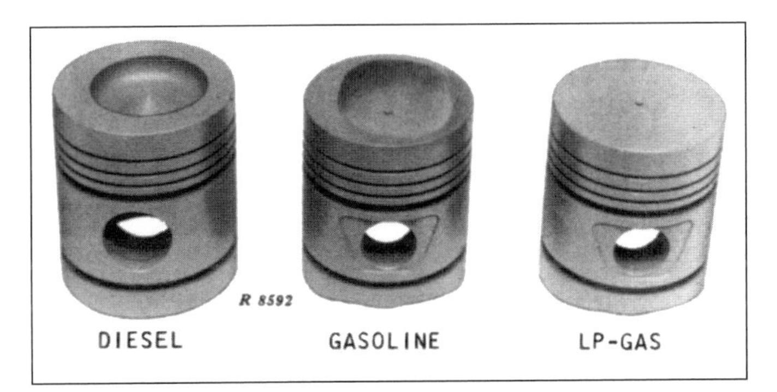

The piston for the 4010 LPG engines had a flat top, unlike the pistons for the gasoline and diesel engines which had cast depressions in the top.

The crankshaft and cylinder block had four main bearings on the six-cylinder engines, and three main bearings on the four-cylinder engines. Since all gasoline and diesel cylinder blocks were machined in the same production equipment, the cylinder bore spacing was common. The common spacing was enough to allow for main bearings on either side of the diesel engine cylinders. The fewer bearings in the spark ignition engines, and the smaller bore diameter, meant there was some extra length in the engine. This penalty was caused by requiring common machining equipment, but it was an acceptable cost compromise.

The combustion chamber in the cylinder head of the spark ignition engines was an oval hat-shaped dome-type chamber. The intake and exhaust valves were seated in the top of the dome. The oval shape was enlarged enough on one side to allow including the inclined spark plug. The LPG engines had a flat top piston, and the gasoline engine pistons had a cast depression in the top. The depression was to lower the gasoline engine compression ratio from that required for LPG combustion.

The combustion chamber design was originally developed in the OX 21 two-cylinder gasoline test engine. Early on it was recognized as being so successful that its features were adapted into the 1956 Model 520, 620, and 720 spark ignition engines.

Diesel Engines

At the time of the work on the experimental OX and OY engines, 1,800 rpm was the maximum rated speed of most competitive open combustion chamber engines. One way of attaining a higher power output at a given cost and package size was to increase the rated speed. Therefore, the 2,200 rpm rated speed for maximum power was chosen. As explained earlier, flexibility was additionally necessary to obtain 2,500 rpm for transport operation.

This higher speed specification challenged what was then known about engines; particularly in the camshaft design, the valve train design, and the fuel injection system. Questions arose on the effect on valve timing, camshaft profile, combustion chamber shape, intake valve and port shape, and the shortened fuel injection period. Since that time, many competitive tractor engines have used higher operating speeds.

Good fuel economy has long been a goal in tractor engine design with Deere & Company. The two-cylinder tractors established low fuel consumption records at the Nebraska Test. It was known early in the design and test program that the new tractors would be more complex and have more power consuming auxiliary systems. Power steering, power brakes, larger hydraulic pumps, transmission synchronizers, and other features would increase the tractor productivity. The added accessories also would increase the parasitic losses between the engine crankshaft and the driving tires. Consequently, fuel consumption reduction was the object of early and continuing work.

Diesel engine studies were first conducted by Deere in the late 1930s and early '40s. These "PX" engines were two-cylinder horizontal designs fitted into modified Model D tractors. An early decision was made to concentrate on the open combustion chamber design vs. the pre-combustion chamber design. The lower fuel consumption and good starting were obvious benefits. This decision was never reconsidered from the Model R tractor onward.

Diesel Fuel Consumption Investigation

The OX 21 engine tests discussed previously were conducted from 1950 to 1953. These tests developed various valve timing, fuel and air supply, combustion, starting, and cooling parameters. Based on the test results, it was decided to make the bore dimension of the engine less than stroke dimension.

Low fuel consumption in an open combustion chamber engine is helped by the fuel burning efficiently as it is injected. A combustion front occurs with fuel injection as the cylinder air temperature ignites the fuel. Combustion is maintained as long as fuel injection continues and fresh air can be supplied. Getting the air to the fuel is helped by imparting a circular motion of the air in the cylinder.

This motion can be made by shaping the intake manifold passage, positioning the intake valve, intake valve period of opening, and other means. On this program a wooden mock-up of the combustion chamber was fitted with an intake valve and a transparent cover on the piston side. Small flags were placed in a circular pattern on the combustion chamber surface. Compressed air was introduced into the intake valve passage. The flag directions were then photographed at various intake valve lifts. By this procedure, the intake port and valve positions were improved. Later, full-scale engine tests provided correlation with these air swirl studies.

In addition, the piston top had a modified toroidal combustion chamber. The shape is similar to a doughnut with about one-third of the top side sliced away. With this chamber, as the piston approaches the top of the stroke, the air is squeezed into a smaller diameter. This technique, called the "squish effect," accelerates the circular air velocity by about 40%. Some experimenting with the diameter of the piston chamber is usually helpful.

Fuel consumption is also improved by tailoring the fuel injection pump and injection nozzle characteristics. (These components are described in a later section.) The configuration of the exhaust system, and exhaust valve lift and timing also affect the fuel consumption, with each component requiring separate individual testing.

After all of that re-designing, the engines had high specific power, at or below 0.4 pounds/brake horsepower-hour fuel consumption (before installation in the tractor), and good cold weather starting. The heat rejection to the cooling system was a low 30 BTU/horsepower/minute. The weight of the six-cylinder engine was competitive with 12 pounds/horsepower and 3-pounds-per-cubic-inch displacement.

The first start-up of a new design engine is a time of tension and excitement for the interested observers. In the late 1940s a new engine was to be started up. The plant manager, chief engineer, and other notables were invited to experience the event. In those days most two-cylinder engine testing was done in a stripped tractor complete with wheels. As the belt drive to start the engine was engaged, Jimmy Parsons, a dynamometer operator, used a 2-pound hammer to hit upon the wheels of the test tractor in his test cell across the aisle. Consternation turned to sheepishness when the notables realized the anticipation they had exhibited.

Diesel Fuel Injection System

The higher operating speeds specified for these engines dictated increased capabilities in the fuel injection system. Governor response, starting, fuel delivery, smoke readings, and other characteristics were factors that were affected by the choice of injection pump and nozzles.

The Roosamaster pump, made by the then Hartford Machine Screw Company, was selected early The pump is now called Stanadyne after the new company name chosen by the manufacturer. This rotary injection distributor-type pump was uniquely different from the previously ubiquitous Bosch multi-plunger style. By 1959 the pump was also manufactured under license by the C.A.V. Company of England. It was then used by the competitive Ford, International Harvester, and Allis-Chalmers tractors.

While the Bosch system had a plunger for each cylinder of the engine, the Roosamaster had a single plunger. The latter featured inlet metering, variable start of injection, and constant cutoff of injection timing. The model chosen was their DB pump which included the fuel transfer pump, the governor, the pumping section, and the drive. The pump was driven off the oil pump drive shaft. This shaft was in turn driven by the camshaft through crossed helical gears. It had the same location and drive arrangement as the distributor on the spark ignition engines.

The inlet metering characteristic provided low internal pump flow control forces. Therefore the governor parts were small and lightweight and permitted governor regulation well within the 8% to 10% speed variation allowed on the engines. Some speed drop or rise is desirable before a governor responds, to decrease the frequency of speed adjustments, thus avoiding a hunting or cyclic characteristic.

The pump was a good choice. It was adapted in succeeding Deere engines with similar good results. The excellent characteristics included: low wear in 4,000 to 5,000 hours with proper filtration system maintenance, lower initial cost, and low diesel detonation noise at slow engine speeds. The lower cost resulted from the single plunger and small lightweight governor parts. The lower noise resulted from the fixed cutoff and variable start of injection. The later start of injection at slow engine speeds lowers the noisy peak combustion pressures because they occur closer to the piston top dead center position.

This view shows the principal parts of the 4010 diesel fuel system.

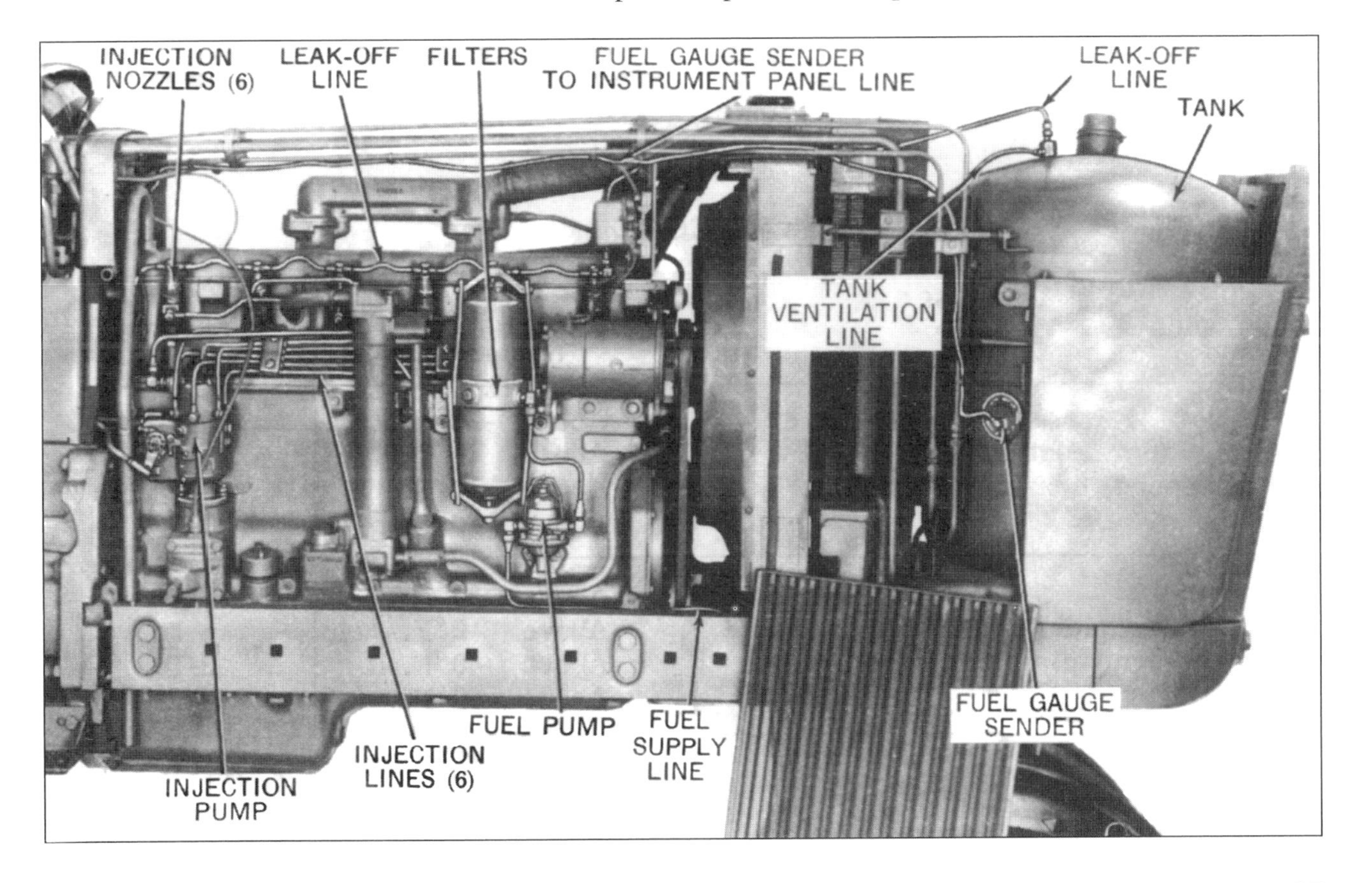

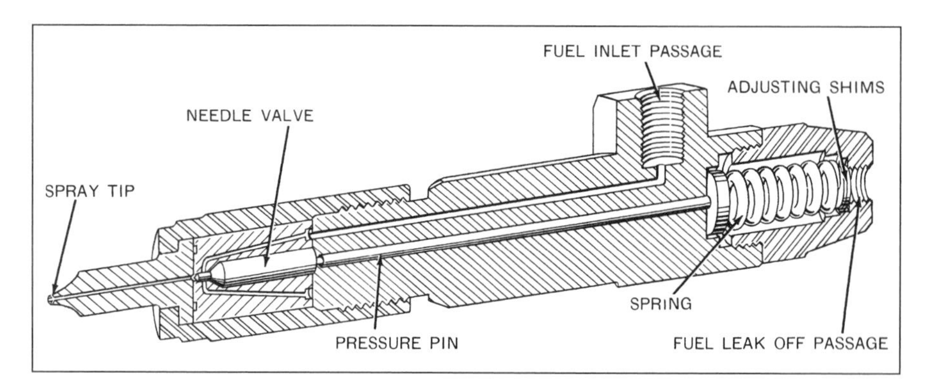

This cross section drawing shows a diesel fuel injection nozzle. The nozzles were tailored to fit the new combustion chambers.

Early tests showed the system was sensitive to flow restriction differences, especially caused by two adjacent nozzles in the firing sequence. The result was an irregular combustion firing characteristic near 1,200 rpm and near high idle of 2,650 rpm. The addition of a delivery (flow check) valve corrected this problem. It was located in the pump rotor shaft in the pump discharge circuit. The pump required only two levers. One was for fuel shut-off that was controlled by a cable to a pull button on the tractor dash. The other was the throttle speed control lever. It had both slow idle (600 rpm) and fast idle (2,650 rpm) adjustments. All other operating characteristics were built into the pump.

The 21-millimeter diameter nozzles were similar to those used on the earlier two-cylinder diesel tractors. Of course, the nozzle tip hole size, number, and angle of spray were tailored to the new combustion chambers. Fuel leak-off from the nozzles and the fuel pump was routed to the fuel tank. This leak-off assisted in cooling the nozzle parts. Over time the fuel tank temperatures would consequently rise as the tank fuel level lowered.

Diesel Engine Design Comments

The engines did not have any external oil lines. However, all fuel conducting lines were external so no leakage could occur to the crankcase. Two thermostats were used on the six cylinder engine's cooling water manifolds to assure good water and steam flow in up and down hill operation.

In conclusion, the comments by the late John Townsend, diesel engine design project engineer, were perceptive:

"No (new) engine can be a scaled up or down version of a successful engine. . . . Very little credit can be taken by a designer for making a good diesel engine. It is a trial and error proposition and good development work to get the best air movement, combustion chamber shape, injection duration, etc. The above items will be changed many times in the (production) life of an engine."

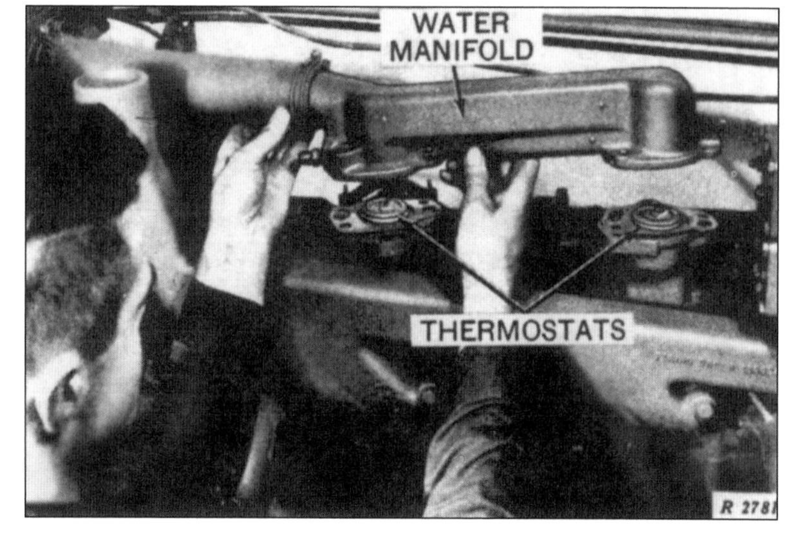

The 4010 water manifold connected the two thermostats to the upper radiator hose.

Transmission, Final Drive, and PTO

Imagine you are a transmission designer and your supervisor came to you and asked you to provide a new design that: (1) would transmit 25% more horsepower than the previous model, (2) would not cost any more to manufacture, (3) would have a smaller overall package size, and (4) could justify the cost of the new manufacturing equipment required for its production. Those were the approximate goals for the OX (4010) transmission.

An earlier chapter described the importance of reducing the transmission length. To reiterate, the concepts used resulted in the speed change and range shift functions being within the same length and compartment. The length saved may be compared to that of having two end-to-end compartments. Conventionally, the final drive reduction was made on two separate horizontally positioned shafts. The planetary final drive that was chosen further saved length over the more conventional spur gear final drive. The planetary final drive also eliminated bending forces on the shafts, which resulted in better tooth contact under load.

Obtaining this arrangement did require adaptation in applying concepts used in other types of transmissions. The concepts were different from those used by the agricultural competition.

Planning the Transmission Types

When the 3010 and 4010 New Generation tractors were introduced, five types of transmissions were planned. They were in various stages of design and testing by Product Engineering:

1. Range-synchronized (the basic)
2. Range-torque shift
3. Range-torque shift with torque converter
4. Full torque shift (eight speeds forward, three reverse)
5. Full torque shift with torque converter

The plan was to use the same factory machine equipment, at least as much as possible, for all five transmissions. Only the range-synchronized transmission was available on the 3010 and 4010 agricultural tractors. In 1963 the full-torque shift (or power shift) design became available in the 3020 and 4020 tractors.

These workers were assembling the 4020 power shift transmission at the Waterloo factory.

Shift timing and sequencing were such that the torque converter was found both unnecessary and undesirable for agricultural use. The torque converter was never released to production for the agricultural tractors. The range-torque shift transmission also never reached production. However, the hydraulically-operated power-shifting clutches designed for the range-torque shift transmission became the front unit of the full-torque shift transmission. The latter is more commonly known as the power shift transmission.

Range Synchronized Transmission

Early in the design process it was decided that the transmission of the 3010 and 4010 would offer range-synchronized shifting. This lower cost approach achieved many of the advantages of powered range shifting. The popularity of this transmission has demonstrated the soundness of this decision.

Practically all the competition at the time had some form of powered range shifting. Although as the two-cylinder tractors continued to gain in popularity in their last decade, they were handicapped by not having a powered range-shifting option. An extensive program was conducted but never implemented to provide a 10-speed transmission with a powered high-low shift.

The basic range-synchronized transmission had all mating gears in constant mesh and no gears were shifted.The engagement of synchronizers provided much lower hand-lever shifting forces than the sliding gear meshes used on the two-cylinder tractors. The range-shift transmission had synchronizers on the high, low, and reverse functions on the input shaft (or drive shaft) to the gearbox. The input shaft meshed with the countershaft in forward gears (high or low), giving the countershaft two speed reductions. The countershaft was located parallel to, and in a triangular spacing from, both the input and output shafts.

This transmission drive shaft sketch shows the high, low, and reverse pinions, and the low and high range synchronizer.

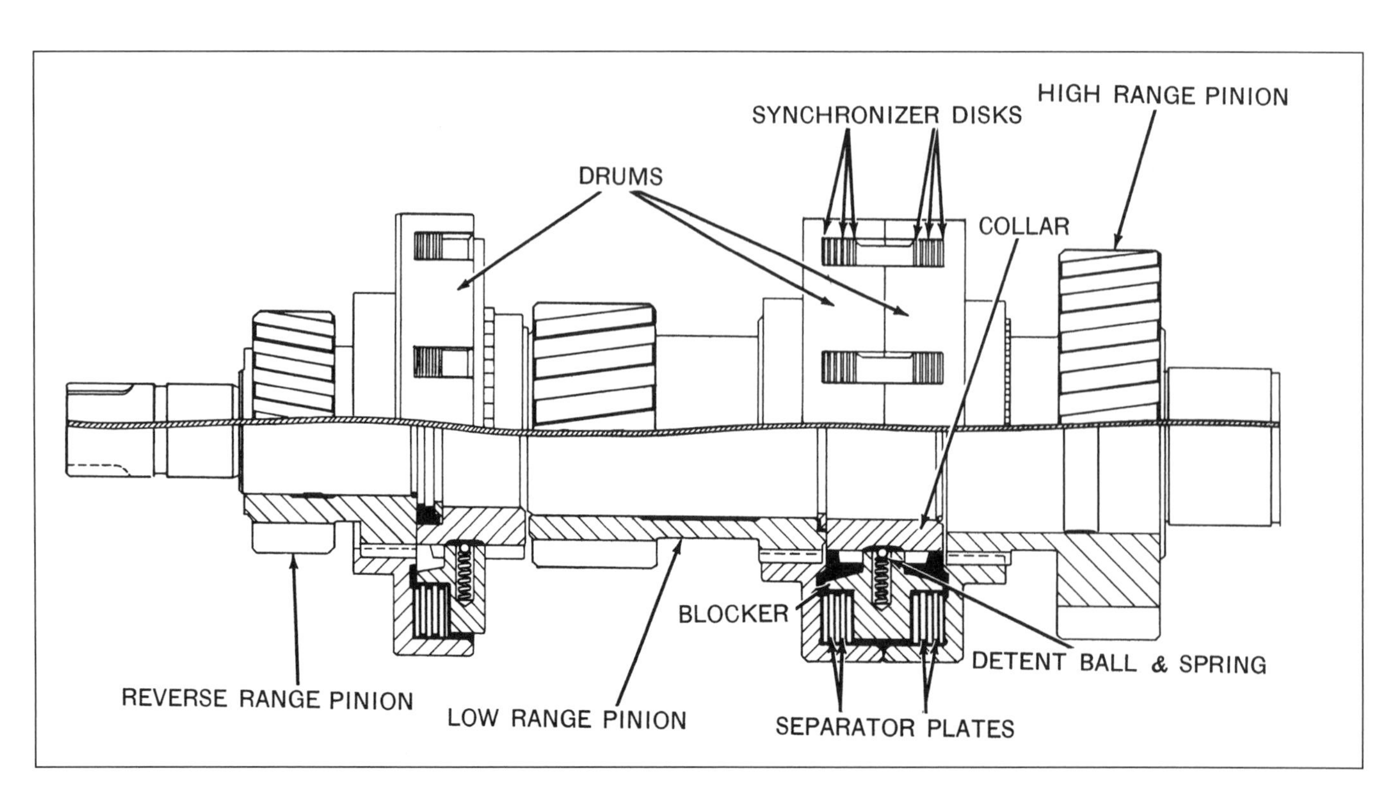

Sometimes the transmission top shaft seizure was bad enough the gear had to be cut off with a torch. Wilbur Davis, Chief Project Field Test Engineer, related such a repair:

"The damaged shaft was taken to a local machine shop for reconditioning as no repair was available. The anxious test crew, including Merlin Hansen, oversaw the turning of the shaft in a lathe to remove the metal transferred from the gear. The shop did not have a grinder attachment for the lathe. So a lot of careful sanding with an emery cloth provided a finish that made a successful repair."

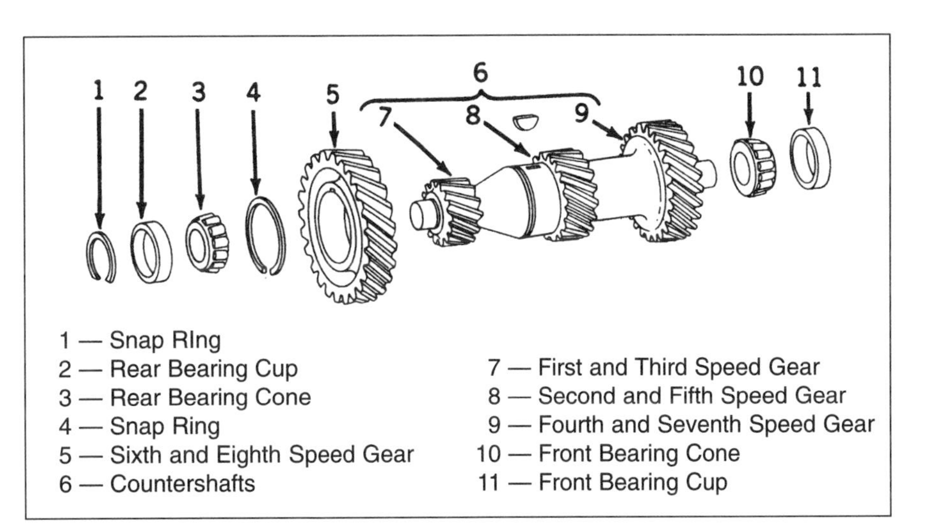

The countershaft was located parallel to the input an output shafts.

The transmission top shaft gears of steel rotated freely on the steel shaft unless held by the synchronizers. This often meant steel was running on steel. Before the required finishes, lubrication, fits, hardness, and clearances were provided, seizures at high speeds occurred.

The four integrally-mounted gears on the countershaft were constantly meshed with four mating gears on the output, or differential drive, shaft. Selective positioning of the output shaft sliding collars gave each of the high and low speeds a further combination of four output speeds, totaling eight forward speeds.

In reverse operation, the rear synchronizer on the input shaft was engaged and the high and low synchronizers were released. Power from the input shaft went through a reverse idler gear on the rear of the output shaft. In the slowest reverse speed, this reverse idler gear was engaged directly by a sliding collar with the output shaft. In the faster reverse speeds, the power went through this same gear with the sliding collar disengaged. The reverse idler gear was always in constant mesh with the countershaft that now rotated in opposite direction from its

This drawing identifies the major parts used on all three transmission shafts.

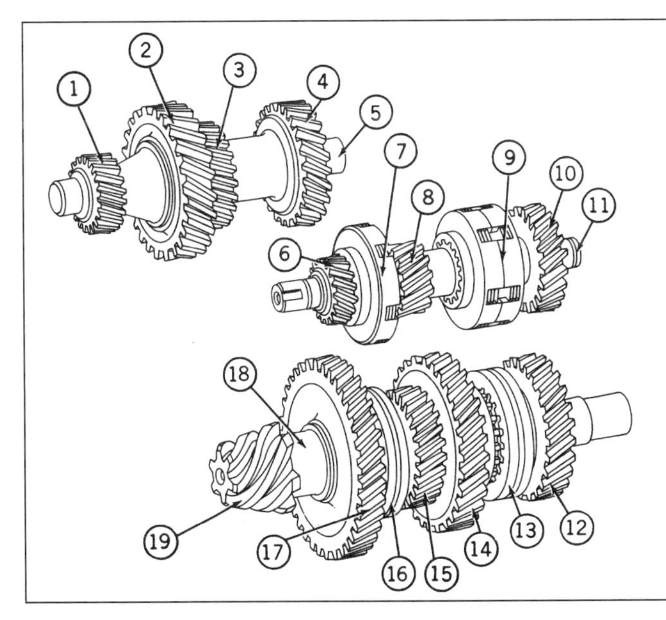

1 — First and Third Speed Countershaft Gear
2 — Sixth and Eighth Speed Countershaft Gear
3 — Second and Fifth Speed Countershaft Gear
4 — Fourth and Seventh Speed Countershaft Gear
5 — Countershaft
6 — Reverse Range Pinion
7 — Reverse Range Synchronizer
8 — Low Range Pinion
9 — Low and High Range Synchronizer
10 — High Range Pinion
11 — Transmission Drive Shaft
12 — Fourth and Seventh Speed Differential Drive Shaft Gear
13 — Transmission Shifter Collar (Front)
14 — Second and Fifth Speed Differential Drive Shaft Gear
15 — Sixth and Eighth Speed Differential Drive Shaft Gear
16 — Transmission Shifter Collar (Rear)
17 — First and Third Differential Drive Shaft Gear
18 — Differential Drive Shaft
19 — Differential Drive Shaft Bevel Pinion

forward operation. Three other reverse speeds could be selected by the output shaft sliding collars. The fastest reverse speed was an unsafe and unusable 15 miles per hour, and therefore was blocked out in the shifting mechanism.

The transmission of the OY (3010) tractor, in keeping with the smaller power and size, was a smaller version of the OX. The machining equipment was similar but separate for the transmission and some of the associated parts. These parts were also used later in the 2510 and 2520 row-crop tractors. In still later years, the power of Waterloo tractors outgrew this smaller transmission and its production was stopped. Newer tractors made in Dubuque and Germany provided the smaller power range.

The use of the countershaft was the key to reducing the transmission length. This transmission arrangement, namely having two reduction steps in the same housing compartment, was accomplished with only 11 transmission gears. The fewer gears provided a less costly arrangement than what was used on many competitive tractors with an equivalent number of travel speeds.

Transmission Gear Design

The transmission gears in the two-cylinder tractors were all spur gears. The number of teeth in contact in the various gear meshes ranged between one and somewhat less than two. Higher gear pitchline velocities permitted more power to be put through the same size gear box. However, experience showed that as the gear contact speeds increased, an increasingly audible and often objectionable noise was encountered. Those sliding gear shifting transmission designs did not permit using helical gearing (where the gear teeth have a short spiral about the gear axis centerline). Modest improvements could be attained as the gear contact ratio approached the relationship of two.

Because of the up to 50% higher gear contact (pitchline) speeds on the OX and OY tractors, it was decided early on that the transmission gears must be the helical design. Aided by helical gearing, the trailing edge of one tooth remained in mesh until after the third tooth in line had begun contacting its mating gear tooth. This arrangement meant that two teeth or more were always in full contact. A helix angle of 29 degrees was used in these gears. Appropriate bearing design and shaft strengths were provided to support the use of the helical gearing. This concept effectively reduced gear noise below the audible level.

Synchronizer Design

Synchronizers had been used in other transmissions for at least 20 years. In most applications the synchronizers speed up engaging gears by about 25% to 50%. In these new tractor transmissions, the speed differential was as much as 211%. Additionally, the demands of tractor clutches required comparatively larger parts than in automotive use.

With tractor transmissions the driven gear is directly meshed with the driving wheels. Therefore, the transmission input shaft and clutch disk must be changed to the gear's speed. These dual requirements of increased speed ratio and speeding up larger masses resulted in a short life for the first-tried cone-type synchronizers. Instead, disk-type syn-

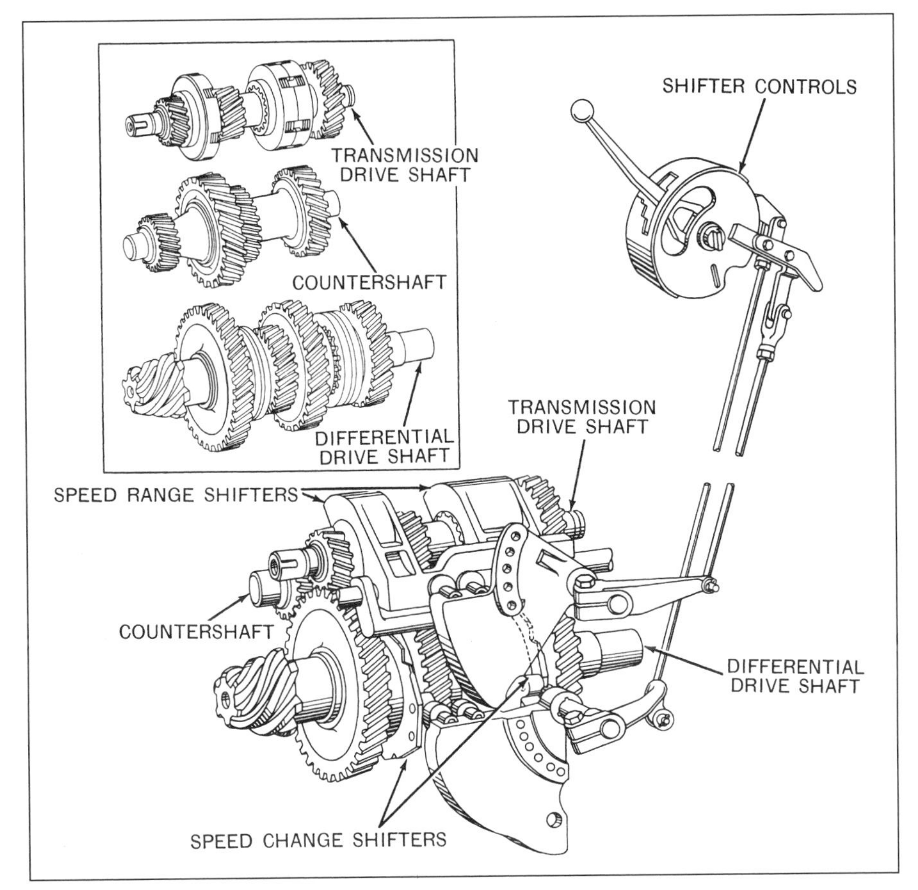

The shifter controls, shifter cams, shifters, and transmission shaft assemblies are shown here in their working relationship.

chronizers, in reality small disk clutches, were designed and developed. These disk-type synchronizers, used with appropriate force on the gear shifting lever, enabled smooth shifts without clashing of teeth, as long as the clutch pedal was fully depressed.

Shifting Mechanism

Each synchronizer and each collar shifter was actuated in its proper sequence by roller cams. The rollers on the shifters engaged cam surfaces milled into castings that rotated in a plane parallel to the gear shafts. These cam castings were mounted on shafts that extended outside the right side of the transmission case. The outer end of the shafts had levers that connected to rods leading up to the dash-mounted shift control lever. The use of synchronizer and sliding collars in the transmission, plus the cam actuation, resulted in acceptably low shifting forces.

Every vehicle must have some secure, safe, and easily applied parking mechanism. On the two-cylinder John Deere tractors the service brakes each had a latch and the operator was expected to reach down (with some difficulty) and place it while the brakes were being applied. Manual shift transmission automobiles and trucks did then, and still do, provide a separate parking brake. The cable operating this brake often corroded in place from disuse. The parking brake continues today even with automatic transmission automobiles.

A survey would probably find that more than 75% of the manual transmission vehicles are parked by leaving them in low or reverse gear. John Deere two-cylinder tractors could be included in those surveys. With this procedure, an unsuspecting operator may start the engine, causing the vehicle to lunge accidentally.

In addition to the parking brake, current automatic transmission vehicles have a parking lock that holds one of the gears in the transmission. With the lock engaged, the engine can safely be started. Sometimes on manual transmission vehicles a safety switch is included requiring the clutch to be depressed before the starting circuit will function.

The OX and OY transmission designers capitalized upon the ease of moving the shift lever for parking. A separate park position was incorporated in the shift quadrant so the engine could be safely started. A small

Incremental reductions in shifting effort were necessary during the design and test program. On one occasion a field test report was returned to Product Engineering Center titled: Instrument Panel Failed. Upon reading further one could find that the operator was frustrated with the hard shifting. He responded by swinging a 2- by 4-inch piece of wood at the gear shift handle. He missed the target, however, and hit the instrument panel.

torsion spring was located at the extreme position on the speed change (lower) cam. When the shift lever was in the park position, this spring engaged the transmission output gears into two speeds at once. This park system was effective, although some difficulty was experienced with releasing it when the tractor was parked on an incline.

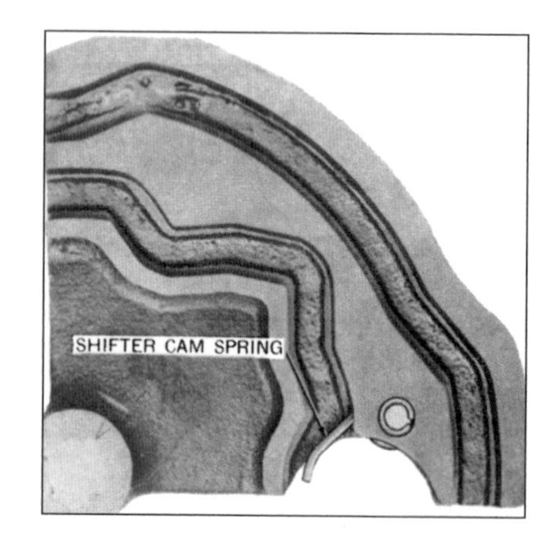

This speed change shifter cam shows the park position spring.

Clutches

Conventional spring-loaded dry disk facings were chosen for the transmission clutch for the main tractor control. These facings were held in the engine flywheel assembly. One friction face contacted a machined surface on the rear of the flywheel.

The main traction clutches were engaged by a pedal acting through a cross shaft and yoke upon a clutch release bearing. This action slightly compressed the clutch apply springs, through appropriate lever linkage, permitting the clutch disk to turn freely of and within the flywheel assembly. The clutch facings were 12 inches outside diameter on the OX, and 11 inches outside diameter on the OY. These sizes gave both tractors approximately the same facing design load pressure of about 25 psi. The clutch capacity was about two times the engine rated output torque. Experience had shown this level of over design was needed to exceed the cyclic torque input from the engine. The capacity was low enough to allow adequate slippage during engagement to make smooth tractor starts. The clutch pedal forces on both tractors were also similar.

The PTO clutch disk assembly was also mounted in the flywheel. When engaged, this clutch disk was loaded by the same springs used on

This sectional view illustrates the transmission clutch assembly,

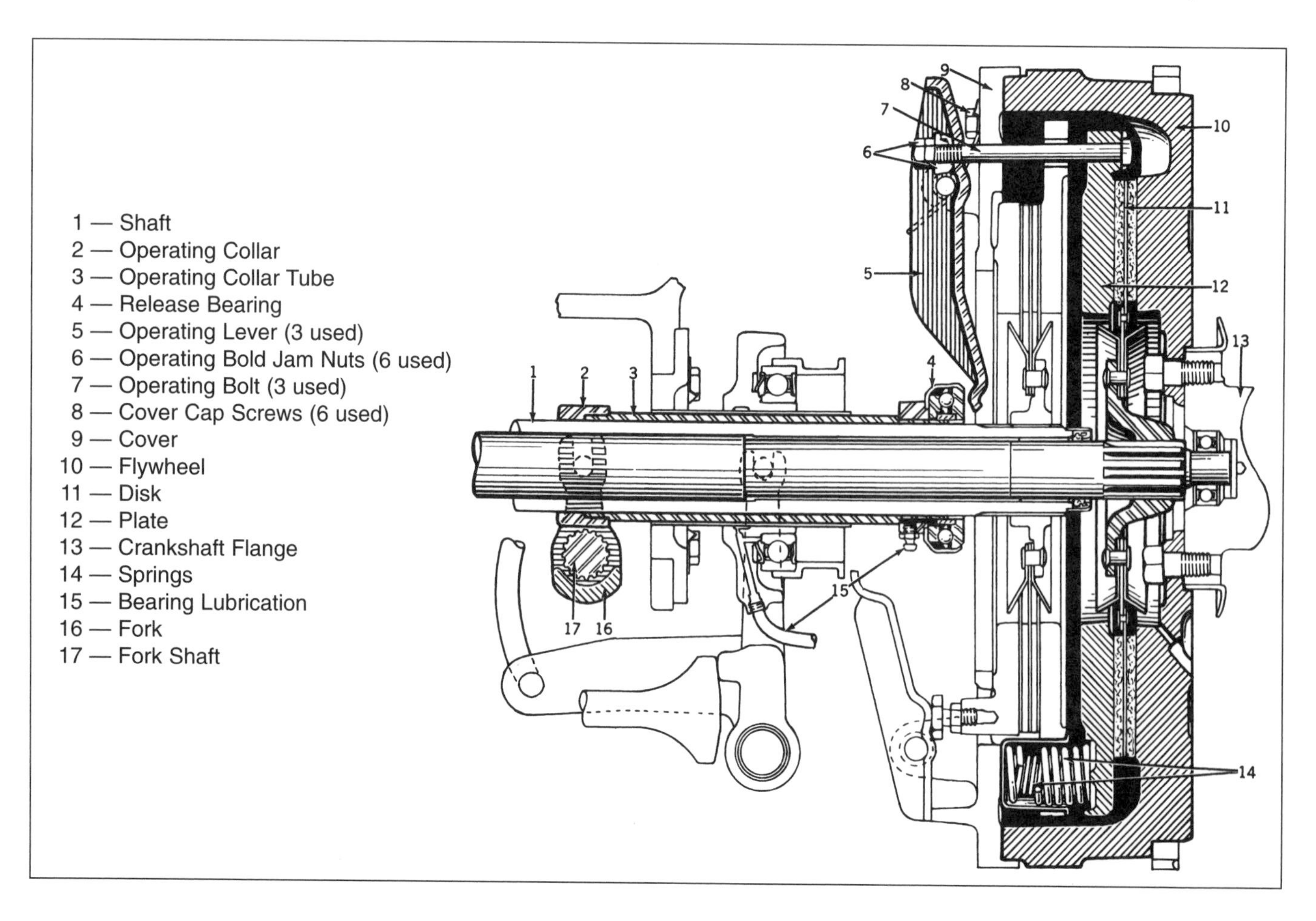

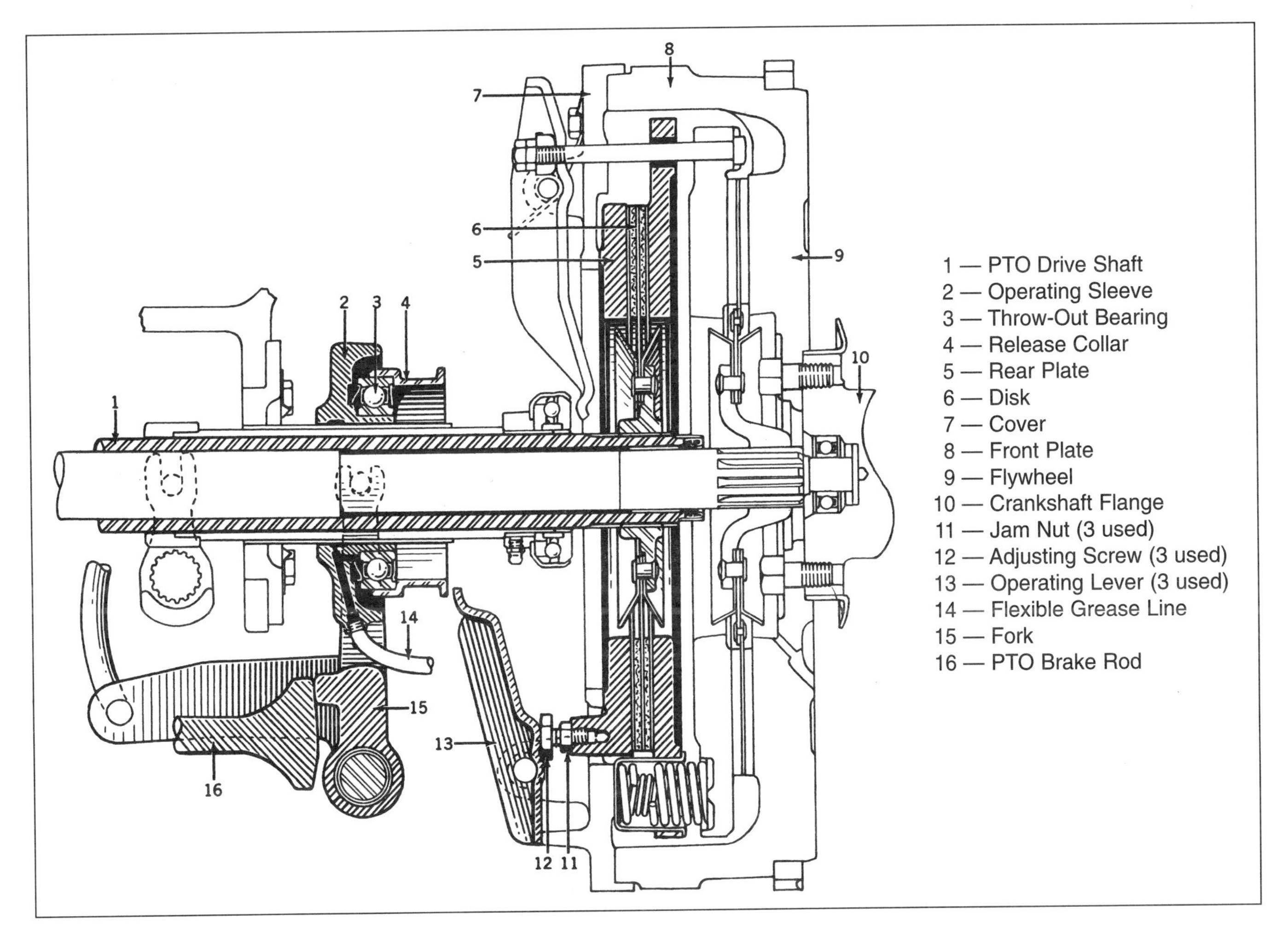

This sectional view illustrates the PTO clutch assembly.

the main traction clutch. Engagement was held by an over-center linkage mechanism which meant the clutch engagement bearing was always loaded when the PTO clutch was applied. This arrangement was acceptable since the PTO usage was expected to be less than that of the main traction drive.

When the PTO clutch was engaged, the slight increase in main clutch spring loading was acceptable and not detrimental. The PTO clutch design load pressure was about 30 psi. The torque capacity was only about 1 3/4 times the engine rated output torque and provided a breakaway feature for the PTO train if some overpowering force stopped the PTO shaft.

The material used in the clutch facings was a composition with embedded brass cloth. The industrial versions used facings of sintered bronze material. Both solid and cushioned clutch facings were tested. The cushioned facings were chosen since they gave improved gradual engagement.

The PTO Gear Train

Beginning in the mid-1930s agricultural tractors have had rotating shafts at the rear end to power drawn or rear-mounted equipment. This feature continues to this day and is expected to be needed in the future. The direct engine-driven construction is most desired. Little application exists for wheel-driven power take-off (PTO) shafts.

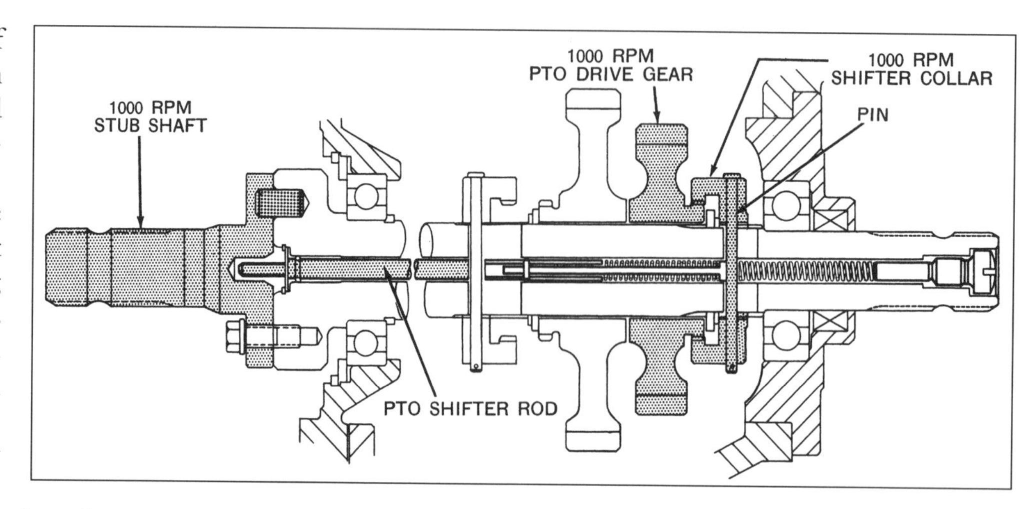

The 4010 PTO shifters are in the 1,000-rpm position.

The PTO direction of rotation and location from the ground, drawbar, and hitch is limited by recognized industry standards. The PTO gear train on the OX and OY had direct routing and gearing because its orientation was recognized early in the transmission design. From the rear of the PTO clutch, four vertically oriented spur gears transmitted the power downward to the PTO shaft. These gears provided a rotational speed of 1,000 rpm at the PTO outlet. Since this speed was a new standard, an optional shifting provision for 540 PTO rpm was made available at extra cost. It was intended for those implements the customer already had which used the slower speed.

The optional 540 PTO speed was obtained in the last two, of four, gears in the vertical train by adding an alternate power path. Selection of one or the other speed was ingeniously provided. Shifting collars were actuated by a shifting rod concentrically contained within the lower output shaft. Shifting was performed by bolting the appropriate PTO spline adapter onto the rear of the shaft. This action prevented an overspeed damage possibility if the 1,000-rpm shaft was used with a 540-rpm implement.

A low-capacity cone-type brake was provided in this gear train to hold the power shaft from rotating when the clutch was disengaged. This brake lacked sufficient capacity to stop any PTO-driven equipment.

A front extension of the power shaft, located about mid-length of the tractor, was a new feature. It powered the front or mid-mounted equipment as needed. Since this new power outlet had no mating for existing implements, only the 1,000-rpm spline was provided. The front of the shaft could turn at either 540 or 1,000 rpm. Since any new implements would be designed for 1,000 rpm, this avoided an overspeed condition. This feature was a situation of anticipating and providing for a greater potential need than ever really arose. While a few implements used this power outlet, it was never very popular. Therefore, the front PTO was eliminated approximately 10 years later when a transmission redesign needed the space.

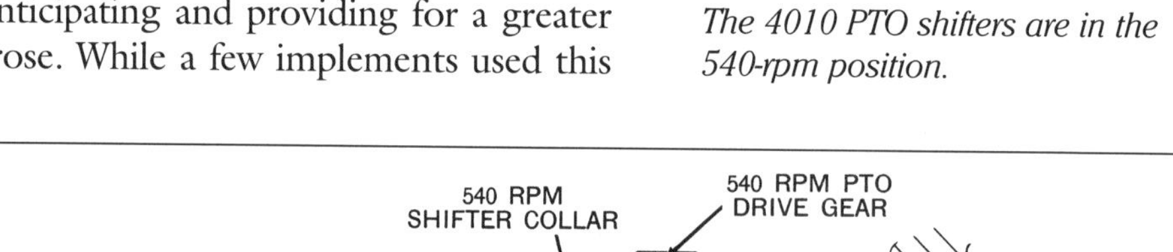

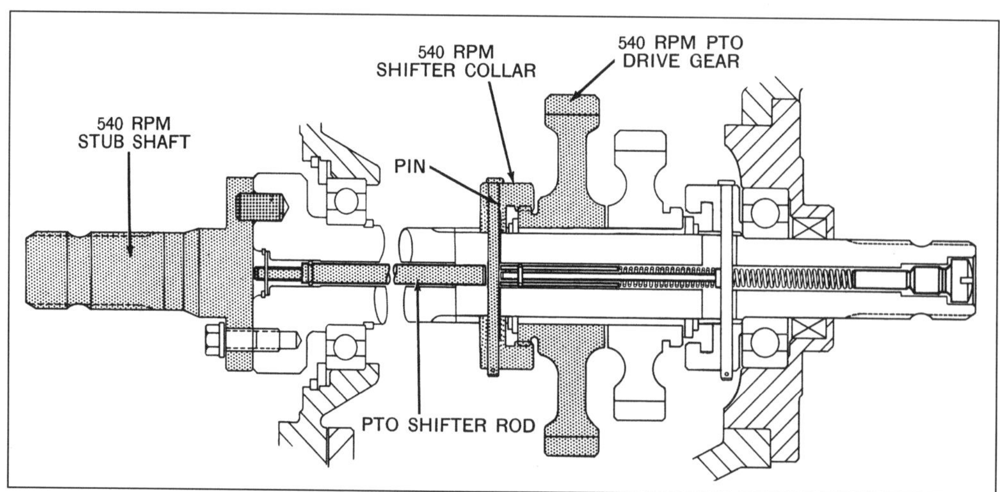

The 4010 PTO shifters are in the 540-rpm position.

Transmission Ratios and Speed Steps

Early tractors had as few as one or two speeds forward: slow and very

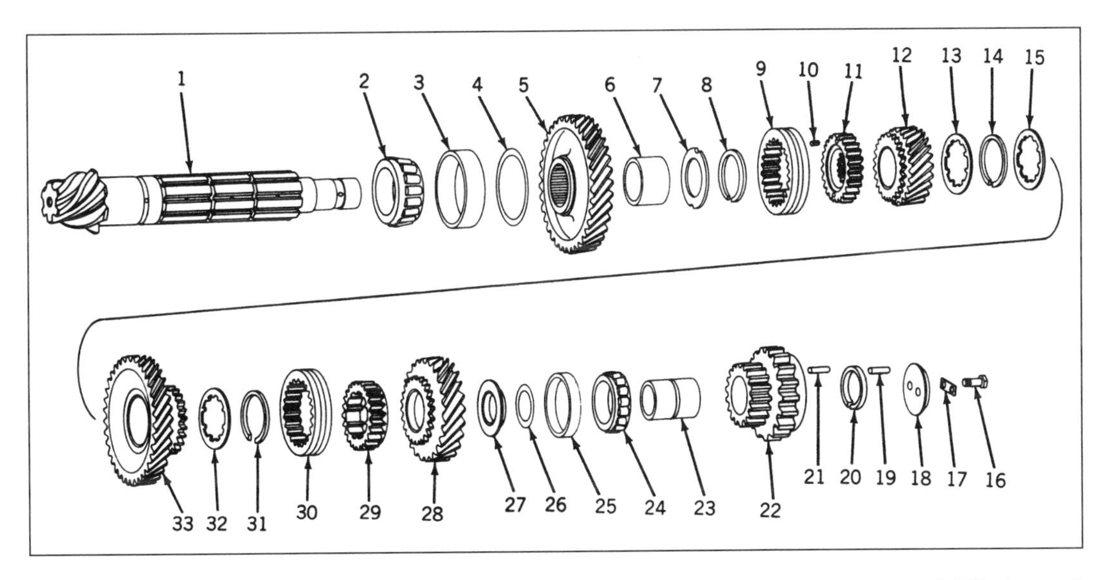

This drawing is an exploded view of the transmission output shaft assembly.

1 — Differential Drive Shaft with Bevel Pinion
2 — Rear Bearing Cone with Rollers
3 — Rear Bearing Cup
4 — Rear Bearing Cup Shim
5 — First and Third Speed Gear
6 — Bushing
7 — Notched Thrust Washer
8 — Snap Ring 0.187-Inch Thick
9 — Transmission Shifter
10 — Roll Pins (2 used)
11 — Rear Transmission Shifter Gear
12 — Sixth and Eighth Speed Gear
13 — Splined Thrust Washer
14 — Snap Ring 0.156-Inch Thick
15 — Splined Thrust Washer
16 — Cap Screws (2 used)
17 — Lock Plate
18 — Plate
19 — Bearing Rollers (33 used)
20 — Bearing Roller Spacer
21 — Bearing Rollers (33 used)
22 — PTO Idler Gear
23 — Inner Bearing Race
24 — Front Bearing Cone with Rollers
25 — Front Bearing Cup
26 — Front Bearing Shim
27 — Fourth and Seventh Speed Gear Retaining Thrust Washer
28 — Fourth and Seventh Speed Gear
29 — Front Transmission Shifter Gear
30 — Transmission Shifter Gear
31 — Snap RIng 0.141-Inch Thick
32 — Splined Thrust Washer
33 — Second and Fifth Speed Gear

slow. They approximated the horse-drawn implement's speed. Their steel lug drive wheels were rough riding, and became rougher riding as the travel speed increased.

When air-inflated rubber tires became available, customers wanted higher top travel speeds than the 6 to 7 miles per hour (mph) usually available. They still wanted the slowest speed in the 1-1/2- to 2-mph range. (Some specialty crops need several speeds even slower than 1 mph.) Today's transmissions commonly range from about 1/0 to 2.0 mph for slowest speed to 15 to 20 mph for the top travel speed. Higher speeds may be reached if a foot accelerator has been provided with the engine throttle control.

This table shows the corresponding transmission speed reductions on the 4010 tractor:

At 2,200 engine rpm, 4010 on 15.5-38 rear tires		
Mph (no-slip)	1.2	16.5
Axle rpm	6.69	92
Overall engine to axle speed reduction	329	23.9
Transmission ratio only (47/1 final drive and spiral bevel reduction)	7	.51

The overall range of transmission reduction of about 13.7 was more than most competition at the time, and provision of the eight speed selections, with an average upshift of about 45% faster than the preceding gear, was notably good. Closer spacings were obtained in the working gear ratios than in other ratios.

In summation, not only was the design compact but it obtained a wide travel span of eight operating speeds. The described good features were not without consequent design iteration. The triangular relationship of the shafts limited the respective ratios available, which complicated obtaining sufficient strength in gear and shaft sizes, and acceptable synchronizer capacity.

Differential and Final Drive

The rear of the transmission output shaft (differential drive shaft) ended with a spiral bevel pinion. It mated with the spiral bevel gear mounted on the differential carrier housing; both of the latter were con-

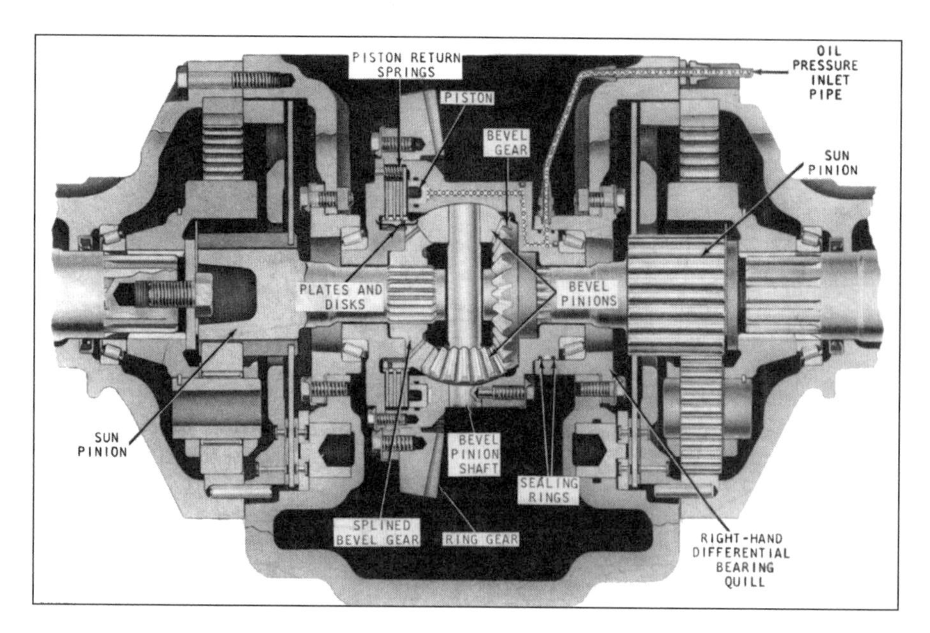

This drawing is a sectional view of the differential and final drive assemblies on 4020 tractors and shows the incorporation of the differential lock assembly.

centric with the axle centerline. A two-pinion differential was used, with the mating side gears driving, through splines, the final drive sun pinions. The latter pinions were floating, meaning that the inner end was located concentrically by the differential side gears and the outer ends by the final drive planet gears. Endwise positioning was obtained between the differential bevel pinion shaft and the axle retaining washer. The inner ends of the sun pinion teeth also carried the brake disks.

The flexibility and adaptability of the 4010 differential arrangement was later put to test. During the 4020 tractor production duration, an optional hydraulically applied differential lock feature was incorporated.

The planetary final drive provided a compact arrangement to achieve the desired final reduction. Some small tractors do not have this added reduction step. As tractors increase in power the cost is lower to make a final reduction step. This permits all preceding parts in the power train to transmit less torque and thereby be smaller. On the 4010 and 3010 tractors the sun pinion is the input, and the planet carrier the final drive output, with the ring gear held fixed. This concept continues with most John Deere tractors today.

At the time of their introduction, these were the only agricultural tractors using planetary-type final reduction. Planetary final drives were in wide use in construction equipment and heavy-duty off-road trucks. Since then, several agricultural tractor competitors have adopted this concept.

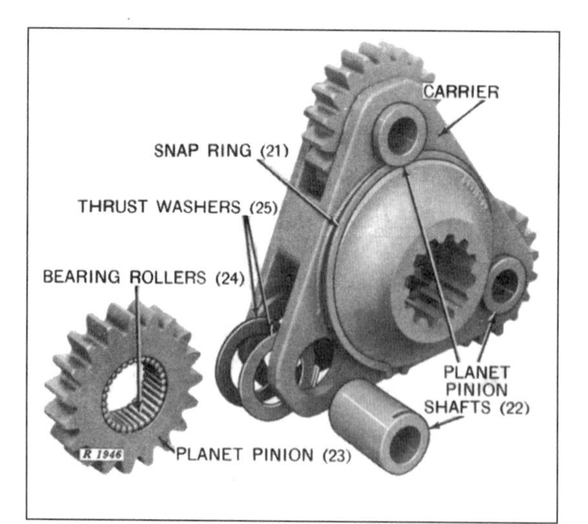

The final drive planet pinion carrier was widely used in construction equipment and off-road trucks, the 3010 and 4010 were the only agricultural tractors using planetary-type final reduction.

The needle roller bearings under the planetary pinions provide an interesting picture of development problems and solutions.

When loaded, one would expect these planet pinion spur gears to have straight tangential forces. The pinion shafts may also be expected to be concentric and parallel the bores in the gears. However under heavy load testing, the early straight needle rollers exhibited galling and spalling of the pinion shafts.

This problem should not arise if the rollers are circulating uniformly around the shaft and giving uniform contact between gear and shaft.

Because this was a problem in many other applications, needle roller manufacturers began offering their product either straight or crowned. With the latter, the roller ends are minutely smaller in diameter than at the roller center. When tried, these crowned rollers gave satisfactory life under load. Evidently, there are small deflections of the planet carrier, or pinion shaft, or inaccuracies in machining, that cause straight rollers to skid. Crowned rollers compensated for the errors.

This translucent drawing shows the location of the hydraulic pumps. Both splash and pressure lubrication were used in this system.

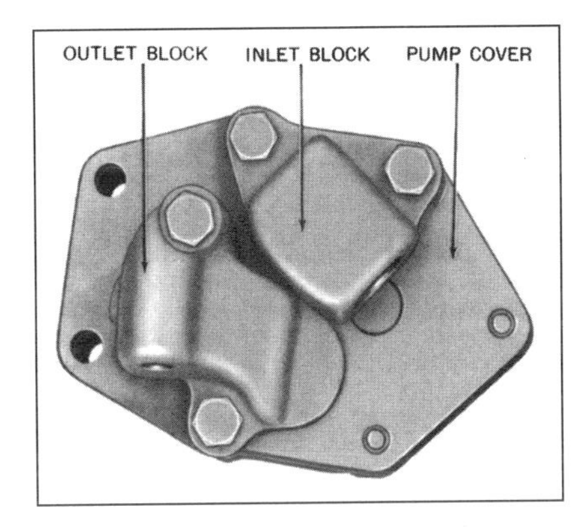

The transmission pump provided an oil surplus over the transmission needs.

Lubrication

The transmission, differential, and final drives of the new tractors relied upon splash lubrication from gears dipping in the oil sump. However, the synchronizers, the constant mesh gears on both the input and output shafts, sliding gear, and collar faces, and part of the PTO train, required pressure lubrication. The lubrication was provided by a small gear pump located at the rear end of the transmission input shaft. This pump always provided oil under pressure while the main traction clutch was engaged, which was most of the time. The pump provided an oil surplus over the transmission needs. The surplus was first routed to charge the main tractor hydraulic pump. Most of the time the main hydraulic pump was at minimum stroke, and did not use all the surplus oil. It was then routed through the transmission and hydraulic cooler. Return (and cooled) oil from the hydraulic system and cooler was lastly returned to the transmission sump.

The transmission case heat radiating surface was smaller in proportion to power on the new tractors than on most other tractors. Also heat was added by the energy losses in the hydraulic system. Hence, the need arose for the oil heat exchanger.

The transmission pump is installed in this picture.

The first experimental tractors had no additional oil heat exchanger. Thus, during early field testing it was temporarily necessary to increase the oil cooling by adding heat exchangers. These were air cooled by electrical fans. Some humiliation occurred because the heat exchangers and fans chosen expeditiously to "save the day" were International Harvester truck cab heaters.

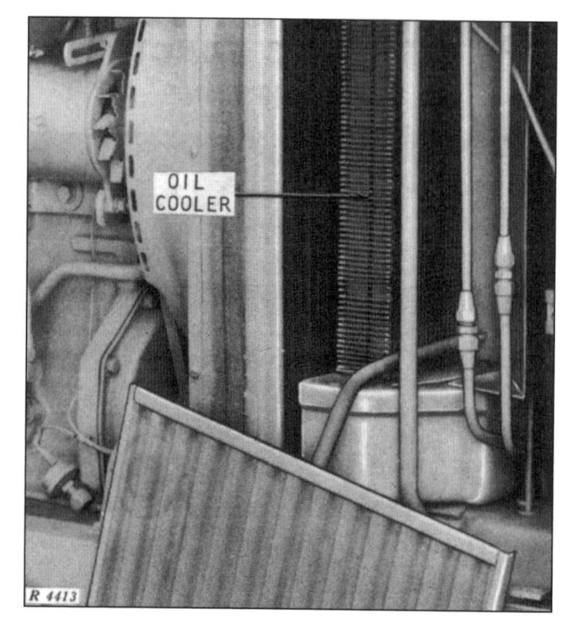

The side grille screen was removed to show the oil cooler location.

Cost

The summary on the cost objective and results are best said by this quotation by Vern Rugen, Chief Transmission Project Engineer (*Product Engineering Seminar 1959*):

> *"Last, but not the least by any means, was the cost. The OX tractor was compared group by group to the 70 and finally the 720 tractor. The total cost of the OX had to be right in line with the 720. — This design of power train — did meet that cost requirement."*

Gear Design Calculations

The gear design calculations were done manually using mechanical Friden calculators which cost about $700. These machines were complex adding machines that multiplied or divided by adding or subtracting the necessary amount of times. No trigonometric or logarithmic functions were available on the machines, so all those calculations were interpolated from standard table values. All calculations, including interpolations, were carried out to eight places. A few newer machines incorporated a square root feature; much needed in some calculations. Otherwise, successive approximate iterations were needed to get square root answers.

Readers today may fail to appreciate the labor required to produce several detailed handwritten pages of calculations on one gearset. Even then, the Friden calculators were a considerable improvement over the all-manual logarithmic calculations previously required. The accuracy and speed of these calculations would have been much faster by the use of the scientific electronic hand-held calculators now available for about $25. And, of course, these calculations would have been easy with the central or desktop computers that have long since become available.

After the gearing was released to the factory for tooling, some of the dimensional data was refined by computer. The first wholly computer-designed John Deere tractor gearing was for the Dubuque-made 2010 tractor high, low, and reverse power shifting transmission. Those calcu-

Even before the temporary heat exchangers were added to the field evaluation tractors, field demonstration situations arose. Wilbur Davis related one such circumstance:

"When the evaluation crew had been at Laredo, some of the senior executives of Deere, including Maurice Fraher and Curt Oheim badly wanted to see the experimental tractor in the field. They were accompanied by Barrett Rich. The test crew considered the hydraulic oil and pump heating and were concerned about the gradual degeneration of the tractor hydraulic performance. They made plans to contend with the problem, yet keep the tractor operating. The solution consisted of several containers of fresh, cool oil.

"Messrs. Fraher and Oheim operated the tractor, and several competitive tractors, with a 4- or 5-bottom intregal moldboard plow. The test crew kept a close eye and ear on the one experimental tractor. Whenever there was a break for coffee, conversation, or for them to exchange tractors, the crew unobtrusively drove the tractor behind the mesquite bushes. There they pulled the

drain plug and replaced with new cool oil. (Author's note: Environmentalists may shudder here.)

"Even when Fraher and Oheim left at the end of the first day, the crew had the tractor in the shop for teardown and replacement of the pump and other parts for the next day. I was invited to, and did, eat dinner with the visitors. The moment I got back to my room, I called Wendell Van Syoc to determine if the tractor was ready for the next day.

"It all turned out well, Oheim and Fraher were satisfied, and the tractor had run successfully. However, we knew we had much work ahead to develop a successful hydraulic system.

"Approximately three years later, while conducting a similar demonstration with the OX, OY, and X77 (2010) tractors, Everett Lee came up to me and said, 'Wilbur, it must be quite a satisfaction to see all the experimental tractors operating without your concern whether they will make the next turn or not.' . . . and it was."

lations were made in the late 1950s on weekends and nights by Harlan Van Gerpen at Moline on the then main frame computer.

Since the early 1940s, most transmission gearing at Deere Waterloo has exploited the principle of "variable center" design. The design made a sturdier gear tooth using a standard gear cutter. By this method several tractor power increases have been accommodated without the need for a transmission space increase. (The technical description of this principle is beyond the scope of this book. Those interested are referred to Earle Buckingham's 1928 treatise: *Spur Gears: Design, Operation, and Production,* published by McGraw Hill.)

If you should sometime see a Deere final drive disassembled, count the gear teeth. Logic says the ring gear should have a number equal to those in the sun pinion, plus those in two planet pinions. Don't be surprised if the ring gear count isn't at least one more tooth than that sum. This results from the variable or spread center design. Additionally, the tooth number choice was made to provide a "hunting tooth" design so the same teeth do not always mesh with every revolution.

Transmission Case

The transmission case on an agricultural tractor is a complex iron casting. It has several varied functions: contain and hold in place the various gears and shafts, bridge the structure between the engine-clutch and the rear axle, provide a radiating surface for dissipating energy losses collected by the lubricating oil, and provide mounting bosses for all the associated lubricating, shifting, final drive, rockshaft, and hydraulic parts.

Early tractors had full length structural steel frames which served to contain the overall stresses and forces a tractor chassis must withstand. They also provided a mounting surface for the tractor's mechanical components.

The skilled designer could lay out a transmission case and surround all the internal parts, provide for mounting all external parts, and achieve adequate casting strength. An obvious decision was the casting wall thickness. A guide in wall thickness was comparison with similar products. During testing the transmission was usually stress-coated and loaded in a laboratory test fixture. This would show the highly stressed surfaces and the case could be redesigned if problems were apparent.

Transmission Oil Development

Provision of a common transmission and hydraulic fluid capable of meeting all its requirements was fundamental in achieving the overall tractor concept. The tractor was made more compact, and less costly, by developing a common oil.

The transmission oil provided several functions:

- transmission gear and bearing lubricant
- transmission pressurized clutch control (power shifting clutches)
- heat conduction
- service brake cooling
- hydraulic system fluid

The first three functions were typically expected of the transmission lubricant. The oil must have ample load carrying capacity to avoid metal-to-metal contact of mating gears and bearing parts. Some energy is given to the transmission oil by the mere splashing agitation of dipping gears and shafts. Parts under load always have some power lost in each gear mesh and each bearing, resulting in heat added to the lubricating oil. This new design added the brake and hydraulic system requirements to the oil cooling requirements.

This heat must be conducted to an external heat dissipating function — usually the transmission case exterior. As stated earlier, on the New Generation tractors air-cooled oil heat exchangers were also necessary. With the heat exchangers, maximum transmission oil temperatures in field operation were about 185° Fahrenheit. The new oil that was developed was an excellent gear lubricant at or below this temperature.

This new oil could have been an SAE 30 oil, except for two requirements. First, the spiral bevel gears required an extreme pressure additive because of their rolling-sliding tooth contact. An SAE 90 gear oil would have satisfactorily filled this requirement, but would have been too viscous (not enough fluidity) for transmission and hydraulic control functions. Second, the oil had to be fluid enough at cold temperatures to flow properly. So the SAE 30 or 90 oil would have required a low temperature chemical additive to reduce the low temperature viscosity.

This oil would then have been satisfactory for use as hydraulic oil, except that care would have to be taken to avoid air entrainment. Sometimes taking the transmission pump suction from a quiet oil location is helpful. Likely, an additive to inhibit foaming also would be required.

One oil candidate used an extreme pressure additive that contained sulphur and chlorine. Moisture is always entrained in some way in transmission oil and during testing this oil the extreme pressure additive would separate. The transmission case insides became coated with the additive precipitate that was slimy and foul smelling. One fortunate day, the need for that additive was eliminated with a material change of the hydraulic pump piston.

The selected oil turned out to be none of the above and was more like automatic transmission fluid. Some of the above mentioned additives are included.

The field test crews also became adept at applying stress coat at the test site, then restressing the tractor to crack the coating. Sometimes transmission case breakage problems were found during testing. Wilbur Davis told of one such circumstance:

"During initial evaluation, a Deere Model 45 front loader was installed and tested on an OX tractor. Mel Long (Test Engineer) came into the office late one afternoon after using the tractor and loader, and said the tractor stopped running. Inspection of the tractor showed it had a unique tilt and it was hemorrhaging oil. Upon closer study we found the transmission case was completely broken.

The loader imposed unique longitudinal forces on the tractor, not previously anticipated. It had literally broken the transmission case in two. Quickly, the transmission case was redesigned and strengthened to withstand those forces."

In another instance, Davis related:

"Early in the program a transmission case had broken at Laredo a few days before a demonstration. The break was high on the rear of the case where the three point hitch upper link attached. There were no spare experimental

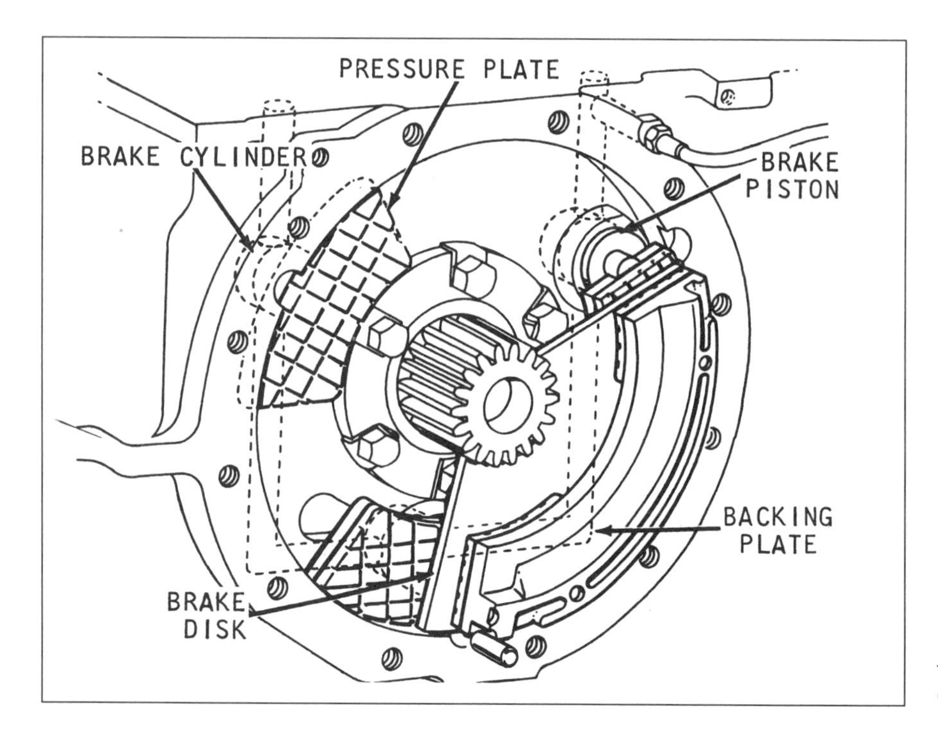

Brake linings, pressure plates, and the backing disk all work together.

cases on the site, and if PEC (Product Engineering Center) had any it was 1,200 miles away. Therefore it was decided to repair the broken case at the test site. Weighing in the decision was the presence of Joe, the welder/mechanic of the permanent test site crew. Joe, of Mexican origin, was known from previous work to possess certain magical skill with a welding rod or torch. Joe and one of the evaluation crew mechanics spent a couple of days welding the patch. Then lead filler was used to provide a flat sealing surface for the rockshaft housing.

"The demonstration called for the tractor to operate with a five-bottom integral plow. Avoiding a failure during that time was best prevented by testing aggressively beforehand. I insisted the tractor and raised plow be taken to the field and driven in fifth gear across a plow furrow. (Davis drove it — he was never one to ask subordinates to place themselves at risk.) The patch failed. So the tractor went back to the shop for two more days where it received a much more significant repair reinforcement. This time it not only passed the bounce test, but also the later driving of company executives."

R. B. Gray in his book "The Agricultural Tractor" stated that during the year 1917: "The Ford Motor Co. of Detroit, Michigan produced for the trade, its first tractor — the Fordson. This was the first tractor to make use of cast-iron unit frame construction and it was not long until practically all of the tractor manufacturers adopted this type of design."

The oil-cooled brake system presented new difficulties. They would squawk and chatter under hard braking. To overcome this problem a "wetting agent" chemical was added to the oil. The brake facing supplier (Raybestos Manhattan) also helped by providing a facing material that had greater difference between the "stick and slip" friction characteristics.

Oil development was first conducted with Lubrizol Corporation and the local Northland Products Company. Robert Giertz led this early work and was able to define the problems and solutions. As the work progressed, Northland saw that the potential production oil volume was larger than their facilities could provide.

With the required specifications in hand, major oil companies were approached and they cooperated. With Lubrizol's continued support in additive formulation, the specification was further refined. Proof tests were established by which other oil companies could also be suppliers, especially in the after-market. The original JDM oil specification has been continuously refined, and with major changes, renamed. This important gain in lubricant development helped competitors improve the redesign of their tractors by using a similar transmission and hydraulic fluid.

The adaptability of tractors to varied usage depends to a large extent upon the versatility of the transmission drive train.

Hydraulic System, Steering, and Brakes

Understanding the reasons for several principal new features in the hydraulic system is very important. Those new features were a closed-center system (vs. open-center), higher operating pressure, and special hydraulic oil.

Closed-Center System

With open-center systems the control valve in the neutral position allows open flow. Conversely, with the closed-center system the control valve in the neutral position is closed allowing no flow. In a closed-center hydraulic system, the pump, acting as a storage tank, continually supplies pressurized fluid to the control valve when it is in the neutral or closed position. The control valve in the closed neutral position does not allow any flow. When the valve is moved, the fluid is directed to the appropriate downstream motor(s) or cylinder(s), and the pump flow is increased to respond. Conversely, the open-center systems, with the control valve in the neutral position, allow open flow. The open-center system had been previously used with the two-cylinder tractors.

The big question is, why change to a closed-center system? Early planning suggested that most of the following tractor functions would be hydraulic powered and be impractical to do with a single pump or a separate pump for each:

- rockshaft for implement lifting
- remote cylinders for implement control
- steering
- brakes
- future fore and aft leveling of hitched implements
- future sideways (lateral) leveling of hitched implements
- future powering of hydraulic rotary motors

Open-Center Hydraulic System

In the simpler open-center system, with the valves in the neutral position, the pump produces a continuous flow that must be returned to the reservoir. For operating one function, this arrangement is satisfactory. However, complications arise if two or more functions must be operated simultaneously. This complication can be avoided with flow-divider valves or special directional-control valves. However, as the number of functions increase, the number of special valves required also must increase. Thus, if providing several functions, a complex system results.

Advantages

The control valve in the open-center system can be designed so the cylinder picks up the load gradually. The peak pressure developed in the system is only enough to move the load. Shock-loading of the system

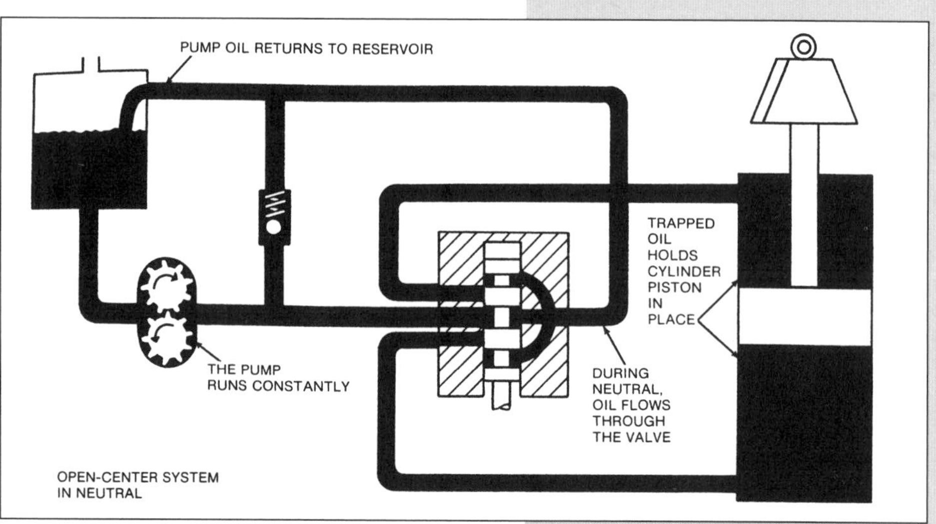

Open-center hydraulic system

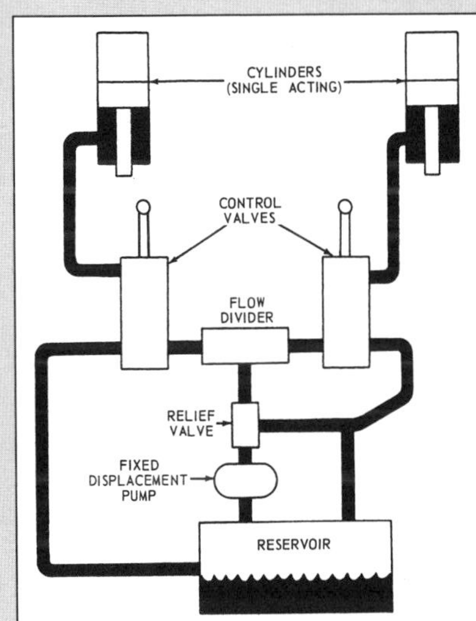

The flow-divider valves helps to avoid problems when multiple functions are being performed in an open-center system.

and parts is also held to a minimum. This reduces wear and breakage of parts, thus reducing maintenance costs.

The control valve also can be arranged to permit operator control of the return flow of oil from single-acting cylinders. Thus, gravity-drop of light implements, or the retraction of loader boom cylinders, can be controlled without compromising the load-raising portion of the cycle.

Limitations

The fit, or clearance, of the spool valve in an open-center system is critical. For example, excess clearance caused by wear results in leakage and loss of efficiency.

The pressure required to lift the load on each cylinder when several cylinders are operated simultaneously must total less than the relief-valve pressure setting. Otherwise, none of the loads may be raised, thus this limits the number of valves and cylinders in such a system.

Closed-Center Hydraulic System

The closed-center system reduces the complexity of the valves required to control many different functions. In the variable-displacement pump arrangement, flow changes to meet the demands of the system. Even when no functions are in use, pressure is maintained. Flow is just enough to make up system leakage, plus whatever cooling flow may be provided.

Pump capacity must be adequate to meet the combined demand created by the simultaneous operation of all the functions on the tractor.

Advantages

The primary advantage of this system is its capability to supply maximum system delivery for sustained periods. Thus, it can provide the capacity to operate front loaders, backhoes, or wheel scrapers requiring comparatively high flow rates for extended periods. The pressure-control characteristics of a variable-displacement pump permit designing the system so it operates over a narrow and nearly constant pressure. Thus, the size of the system actuators can be matched to this pressure. The wall thickness and other physical-strength dimensions of components are also matched to this pressure.

Limitations

The primary disadvantage of this system is economic rather than technical. Variable-displacement pumps for tractor hydraulic systems are expensive.

Pumps of variable displacement were available for many years before their use in tractor hydraulic systems. Their performance over the speed range of tractor engines, coupled with their relatively high cost, discouraged designing a hydraulic system around this pump.

To produce the pumps now being used in farm tractors required a long-term program of design, development, and testing. The result was a pump design having the desired performance characteristics at a cost practical for a farm tractor.

A secondary limitation is the tendency, in some instances, for temperature to rise in the pump. The low flow in standby condition provides less chance to distribute the heat throughout the system for maximum cooling. However, in practice it is possible to overcome this by deliberately providing circulating oil for cooling the pump itself (Fundamentals of Machine Operation, Deere & Co.).

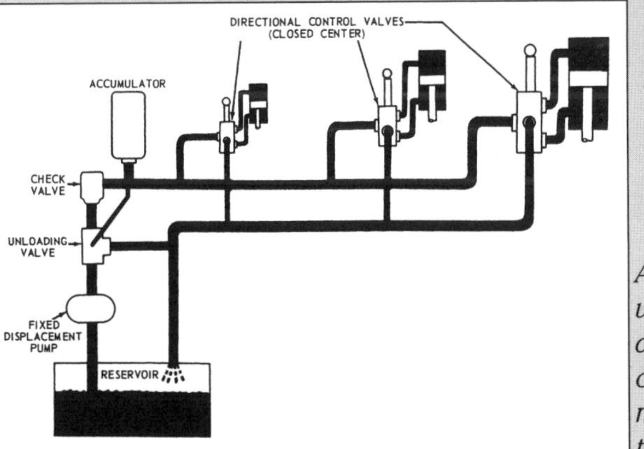

A closed-center system with variable displacement pump changes the flow to meet the demands of the system.

The 4010 hydraulic system components are highlighted on this phantom view of a tractor.

With an open-center system a pump would have been required for each function. Also, if one pump were to move multiple remote cylinders, the flow would first go to the cylinder with the least resistance. Moving all cylinders together was unlikely, so they instead would move in sequence.

With a closed-center system, all valves could be opened and all would move. The pump output needed to be more than the demand. Appropriate flow control valves also needed to be applied.

One central pump and its attendant hydraulic lines was recognized as more cost and space-effective than having multiple pumps with their respective drive systems.

Closed-center hydraulic systems in the 1950s were common in aircraft and machine tool products. They were not used in agricultural equipment.

The disadvantages were expected to be: (1) Early acceptance would be slow because it was new and different. (2) The pump horsepower at high or low system pressure was nearly the same. Although the solution to this was variable pressure control by sensors and feedback lines from each operating function, this complication was not thought to be cost

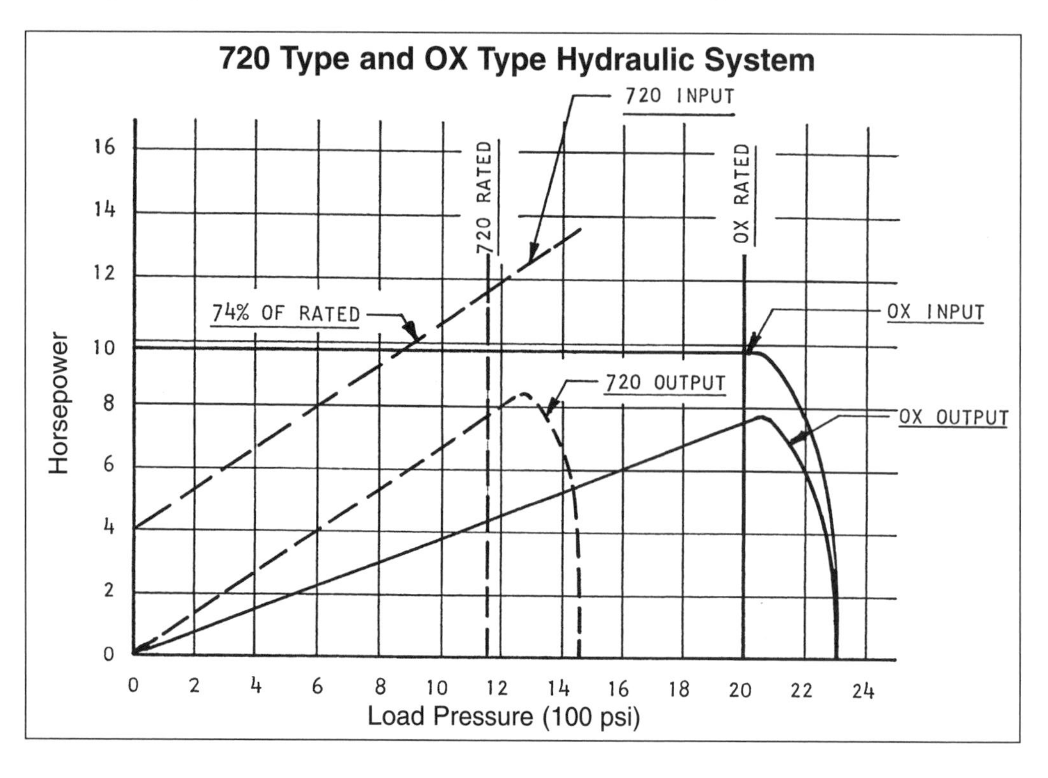

This chart graphs the comparison of horsepower for the closed and open center systems operating with variable load at full flow.

effective. At least one competitor, Allis-Chalmers, later introduced this feature with hydraulic sensing lines. Deere has not introduced this feature, although the company does have the early patents on the concept.

Higher Operating Pressures

The use of hydraulic power to increase tractor control and strength capabilities began in the 1930s:

- 1933 — the first hydraulic rockshaft for mounted implement raising (Deere Model "A")
- 1943 — remote hydraulic cylinders for drawn or mounted implement control (Deere Powr-Trol).
- 1953 — the first factory-available hydraulic power steering (Deere Model 50, 60, and 70 tractors)

In 1949 Deere engineers studied all the industry's tractors and found a correlation between drawbar horsepower, rockshaft lift time, and hydraulic horsepower. At the time the largest tractor was only 40 drawbar horsepower. Another study found that the percentage of hydraulic horsepower to tractor power was increasing with time:

1949 — 13%
1952 — 15%
1959 — 17%
1959 — Industrial wheel tractors 25-50%
OX (4010) — 30%
OY (3010) — 50% (same pump as on OX)

The higher OX and OY ratios were thought important to provide faster implement lift time. They also made the system adaptable to industrial and construction use. This also provided an allowance for the multiple tractor functions being hydraulically powered.

Some competitors were beginning to use increased pressures. It was recognized early in the OX and OY design that these increasing hydraulic power requirements could make the size larger and cost greater of all

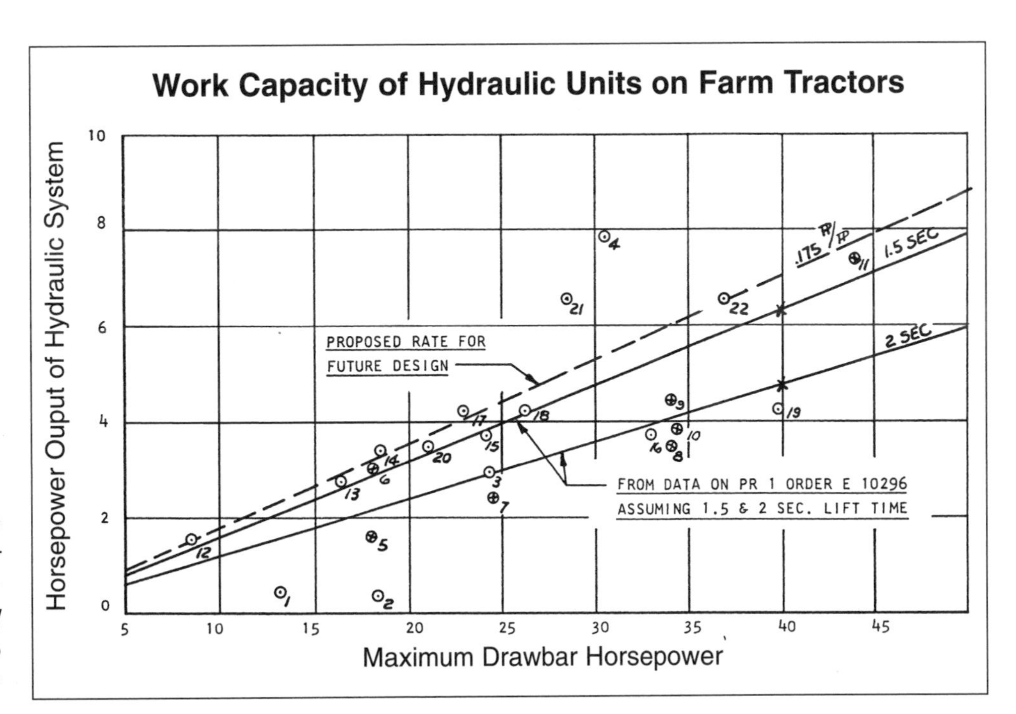

This graph shows the correlation between tractor drawbar horsepower, rockshaft lift time, and hydraulic system pump power.

hydraulic components. The hydraulic system costs would then be proportionally greater than objectives for the rest of the tractor. Therefore, a near doubling of the system pressure, from 1,000-1,250 psi to 2,000-2,200 psi, was decided. The disadvantages included: (1) The incompatibility with existing mounted or drawn equipment systems. (2) The lower cost gear pumps would have excessive leakage and distortion and would be unsuitable. (3) More precise control valves were required to lessen the leakage recognized with the usual spool valves.

With great cooperation from the John Deere dealers, each of these disadvantages was addressed for better solutions to their problems:

1. Education was the principal effort to acquaint the customer with ruptures possible by using older, lower pressure remote cylinders with the new tractors. The practice of shipping new remote cylinder(s) with the tractor was effectively continued, and field conversion bundles were also available.
2. Improved styles of gear pumps were studied. It became apparent that a piston-type pump would most likely have minimal leakage under the higher pressure. Precedence for this existed in the aircraft and machine tool industries.
3. Poppet-type valves, instead of spool valves for reducing the leakage due to higher pressure, were used throughout the system.

Possibly no feature of the 3010 and 4010 tractors caused more controversy among users than the hydraulic system. As much as 15 years later inquiries were still being received. Some consumers questioned why their old (low pressure) remote cylinders would burst. Others wondered why their drawn or mounted equipment circuit caused the oil to overheat. The selection of rotary hydraulic motors for best speed and power output caused a particular difficulty. It was typically a problem with small manufacturers of agricultural equipment.

Special Hydraulic Oil

The need for special oil lay more heavily upon the friction stability of the brakes and the transmission. Designing the tractor to be compact did not allow space for a separate hydraulic oil reservoir. Using the transmission sump to serve dually for transmission and hydraulic oil storage provided a solution. However, it also presented a significant problem.

Since this new oil was used throughout the hydraulic system, the problem was the foreign oil contamination potential that existed. Most problematic was the SAE 30 engine or SAE 90 gear oil used commonly in older systems. Remote cylinders filled with common oil allowed oil intermixing and degrading of the special oil additives. Here again, user education was the principal solution.

Overcoming this problem needed a considerable advancement in filtering. The major achievement was the filtering of all the oil going to the precision hydraulic components.

Main System Components

The major components in the hydraulic system were the pump, heat exchanger (cooler), oil lines, charge pump, rockshaft, selective control, steering, brakes, and sump storage.

Two most unusual experiences occurred during the development of the hydraulic pump; one was with a $3,000 radial piston pump that lasted three hours. The other was with a company that did not want to give a price quote on their vane pump. They believed that in meeting Deere cost objectives they might be asked to lower the price to all other informed customers.

Hydraulic Pump

Some other makes and types of piston pumps were bought, studied, designed, built, and tested. Criteria for a satisfactory pump included:

- Ease of manufacture to obtain the desired dimensional and leakage control.
- Variable displacement, with a rapid means of going from no flow to maximum flow and return. This was necessary to provide prompt response of the hydraulic system functions.
- Cost.

The pump chosen and used during most of the tractor testing was the 8-radial piston pump. The closed end of this pump's pistons were spring loaded against a central cam race. The cam race was mounted with needle rollers on the eccentric cam. For full piston stroke, the camshaft compartment pressure was a minimum. For partial piston stroke, the pistons were held off contact with the cam race by pressurizing the camshaft compartment.

Pressurizing the camshaft compartment was accomplished with high pressure output oil. A momentary increase in the system pressure caused pump control oil to be admitted to the camshaft compartment. This increase caused the pump stroke to be lessened. A quick dump valve was developed to rapidly lower the compartment control pressure and thereby rapidly allow full piston stroke. All the required control was contained in the "stroke control" valve built onto the pump casting.

The pump was driven through a flexible coupling off the front of the engine crankshaft. The flexible coupling helped overcome the torque variations between the engine and the pump. Considerable development

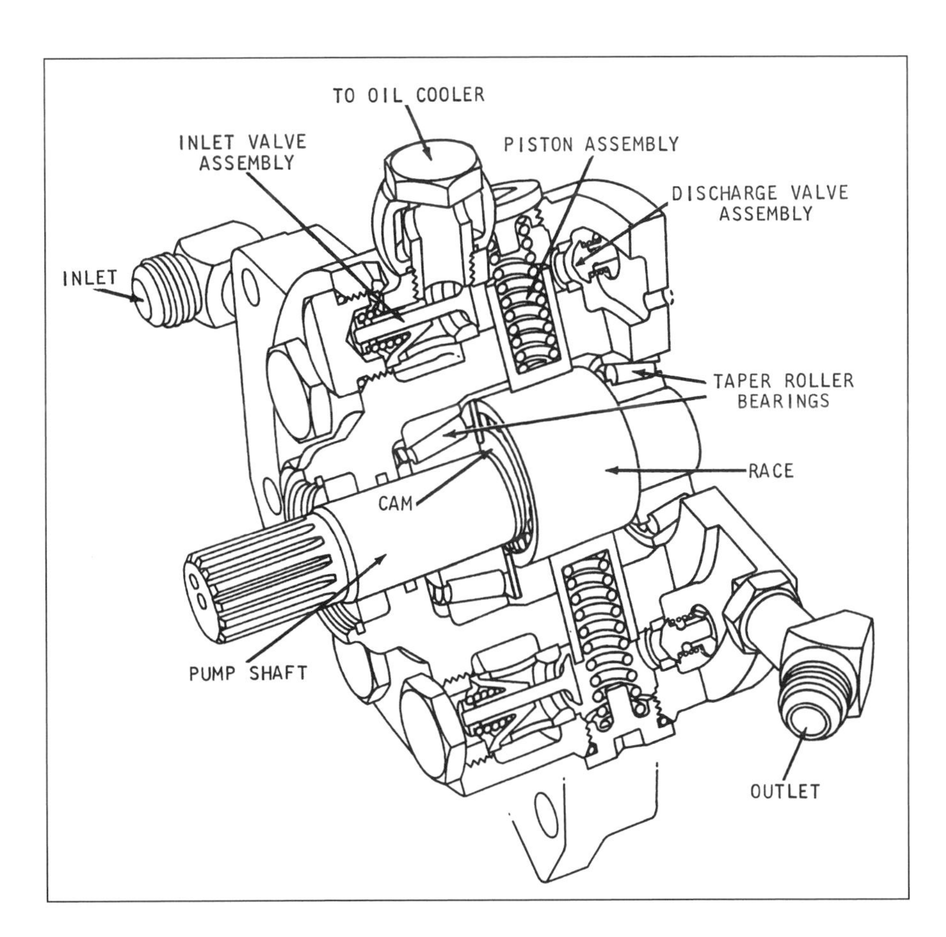

This sectional view shows the production 8-piston hydraulic pump. This pump was used during most of the tractor testing.

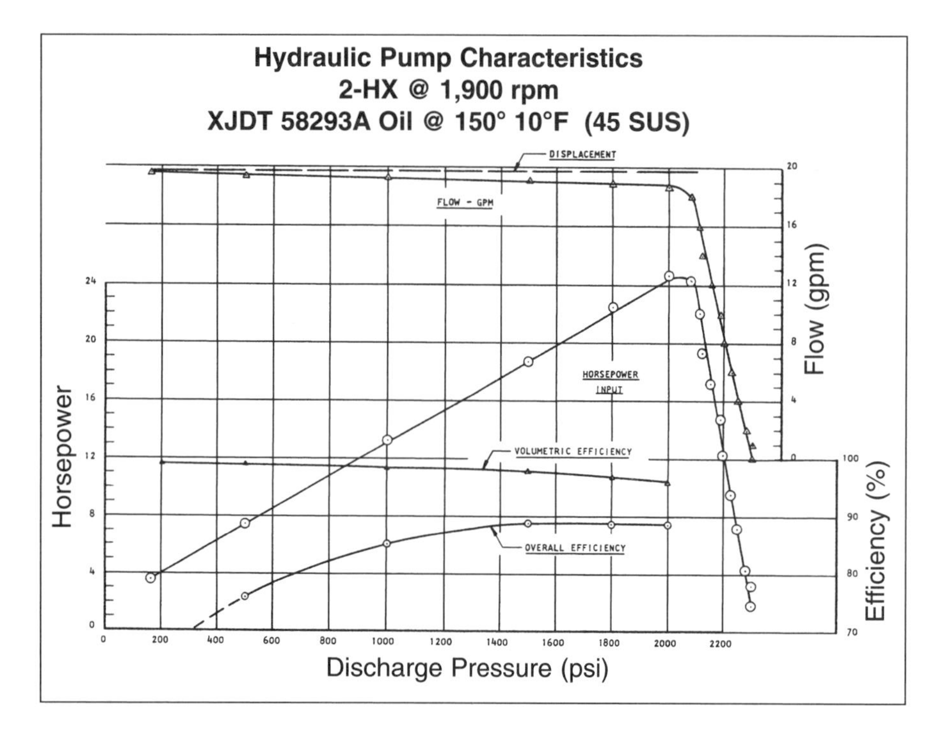

This graph depicts typical 8-piston pump performance.

of the flexible coupling was necessary to achieve the desired level of durability. The flexible coupling also reduced the criticality of the pump to engine alignment. In addition, the pump was precisely mounted off the front of the engine for drive alignment.

One reason for mounting the pump ahead of the crankshaft was the resulting placement and proximity to the hydraulic oil cooler. When the 18-gallon-per-minute (gpm) main hydraulic pump was in full stroke, the 6-gpm transmission charge pump could not provide enough oil. Assisting was the vertical reservoir of oil contained in the oil cooler. The rockshaft (a big demand hydraulic user) could complete its full stroke in 2 seconds. This feat meant about a 1/2 gallon of oil was supplied by the pump and was momentarily drawn from this low restriction oil cooler source. This oil was then delayed in its return to the transmission sump. The oil was replaced in about 6 seconds of the charge pump's output.

*These two graphs show the 8-piston pump response **with** the quick dump of camshaft compartment oil.*

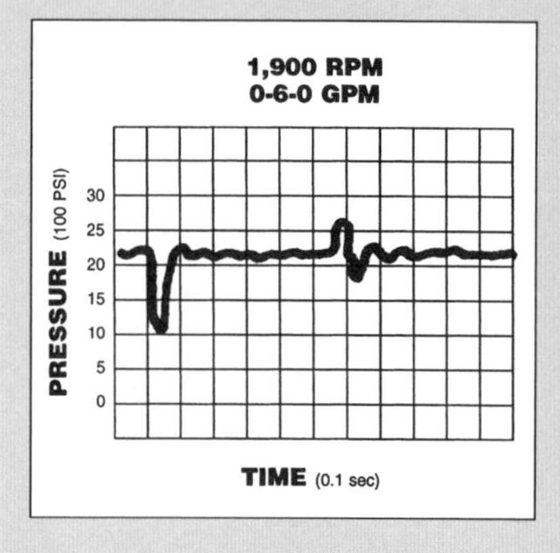

Low flow

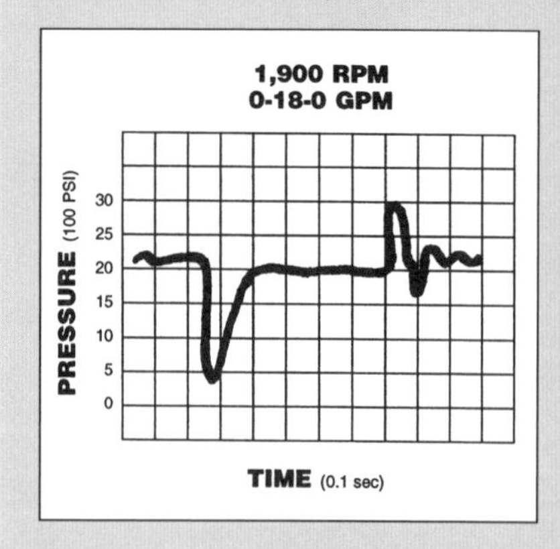

High flow

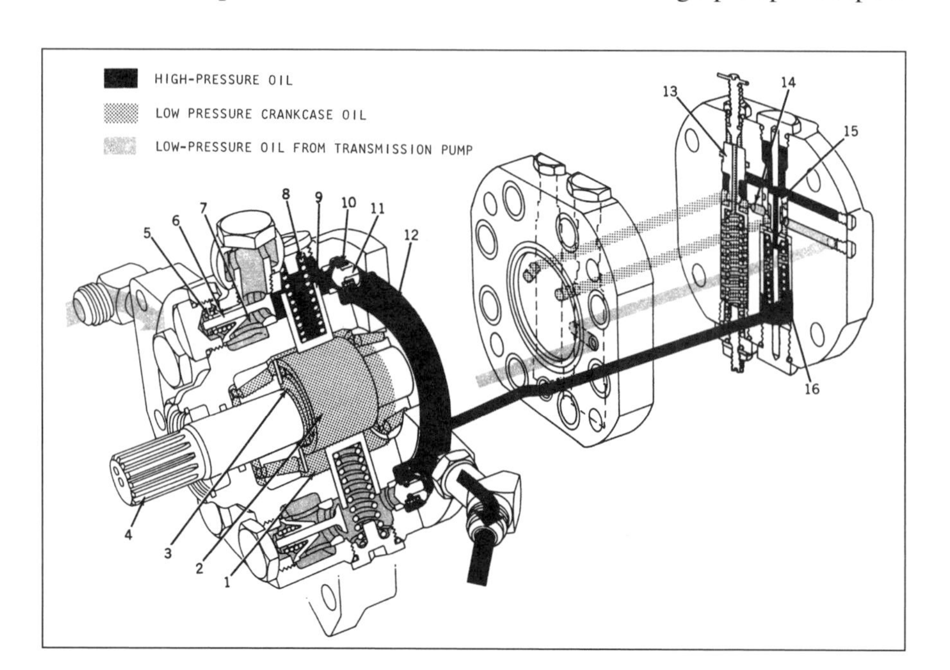

The hydraulic pump circuits are highlighted on this phantom view of the 8-piston pump.

This schematic diagram shows the tractor hydraulic system. The differential lock valve is included as it was used on the 3020 and 4020 tractors.

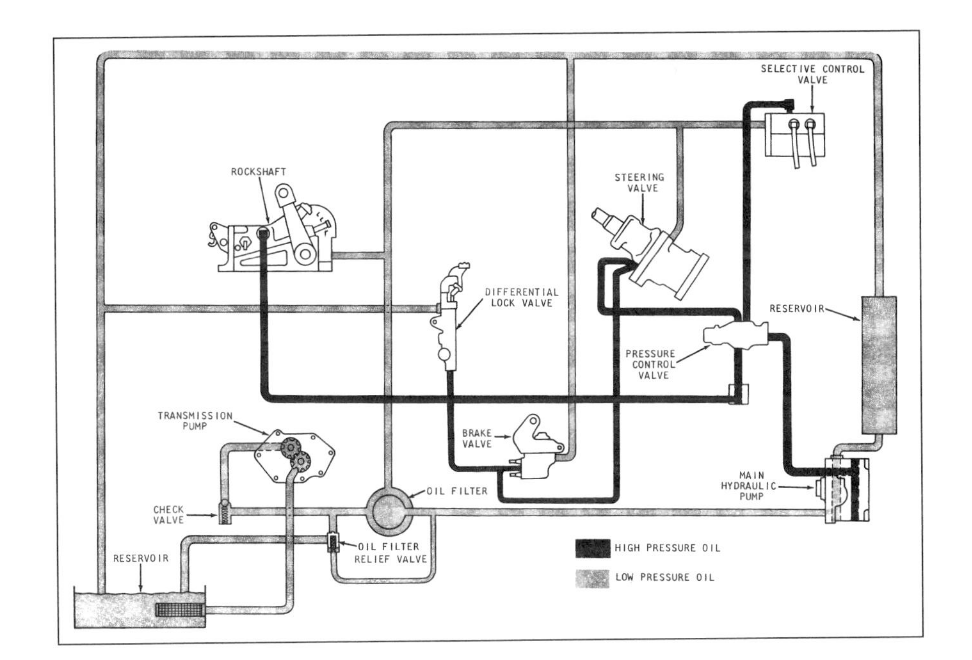

These two graphs show the 8-piston pump response ***without*** *a quick dump of the camshaft compartment oil.*

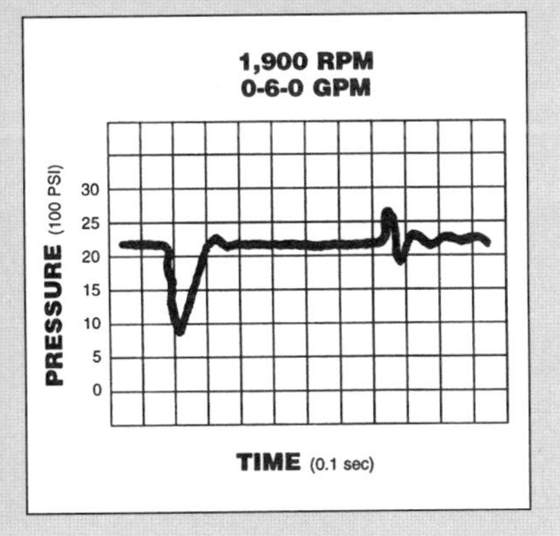

Low Flow

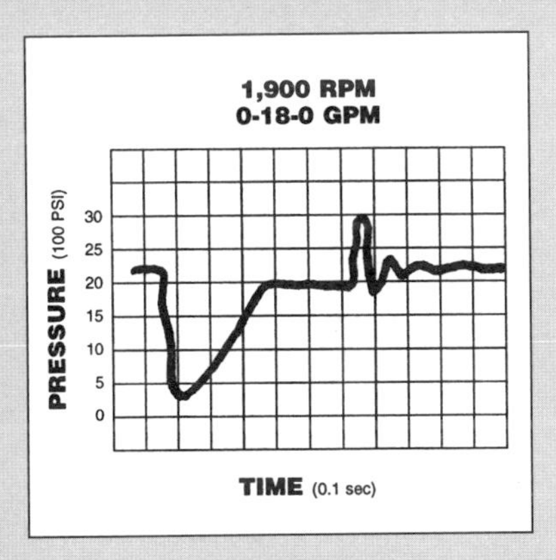

High flow

After the rockshaft cycle, the transmission oil sump was temporarily a 1/2 gallon (of 20-gallon sump capacity) low.

The camshaft compartment pressurization of the pump pistons to control piston stroke was patented by a Wisconsin firm. A royalty of $1 per tractor was paid to that company for several years.

Considerable testing and development ensued to provide durable piston faces, cam face, needle roller life, and sound pump castings. At one time the cam face was unsuccessfully octagonal in cross-section to give a flat face for each piston.

The compressed volume above (radially outward from) the piston was held as small as possible. The pump pistons were held against the cam face by return springs. The low restriction suction and discharge valves were pressure operated, with light assist springs. This pump and its associated close tolerance valving required the factory to institute improved levels of machining and dirt control.

Today, it is hard to realize that in the 1950s the available electrical controls were costly, massive, and unreliable in a tractor environment. Electronic systems as known today were not available. Today, fast accurate response times, rapid proportioning of signals, and reasonable cost are possible using electronics. Therefore, in the 1950s, the OX and OY programs' hydraulic systems required that hydro-mechanical control means be used.

The first hydraulic pump design used a 9-piston pump. This design had high internal leakage and lower efficiency than the previously described 8-piston pump. The lower efficiency generated heat losses that were mostly added to the hydraulic oil. Consequently, several instances occurred where system components also became over-heated from the excessively hot oil.

Control Valves

All of the control valves used poppet-type valves to meter and direct the oil pressure and flow. Movement of the "metering shaft" unseats the ball check valve, thereby allowing the high pressure oil from the larger circumferential groove to pass to the small end of the "outer valve." This

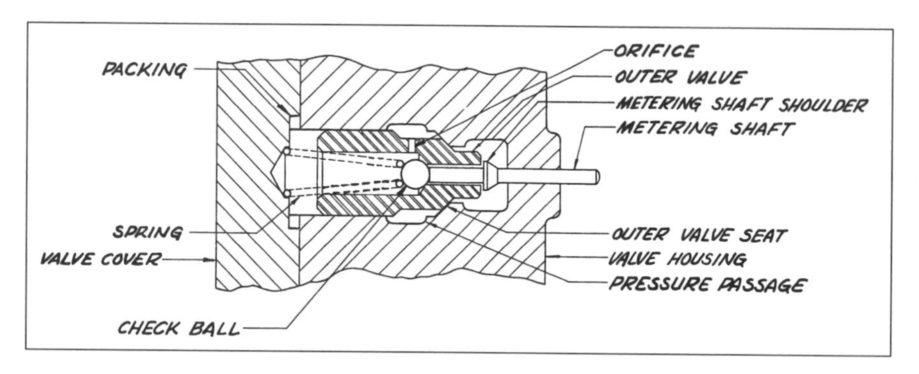

Poppet valves, such as this one, were required for both the outgoing and return flow.

arrangement became the low-flow metering phase of the control valve and nearly equalized pressure on the outer valve so further movement of the metering shaft required less force to lift the outer valve off its seat. The last movement provided the high-flow phase of the control valve and existed as long as the metering shaft was depressed.

When the metering shaft was allowed to return by the control lever, the spring reset the check ball and pushed the metering shaft. This movement began reseating the outer valve because of a pressure differential being re-established.

The sealing surfaces were more easily machined with tapered valve seats that were less affected by wear than the sealing surfaces in a spool-type valve. Also, on double-acting functions, such as remote cylinders, both the outgoing and return flow required poppet valves. An adjustment procedure provided the timing of the inlet and return valve closing relationship.

Selective Control Valves

In the selective control valve assemblies (for implement control functions), the operational modes were slow raise or lower, fast raise or lower, and float. The valves were controlled with cams between the lever and the metering shaft. The spring-centered valves were held in fast raise or lower with automatic return to neutral when the remote cylinder stroke was completed. The float position was beyond the fast lower position and there was a detent to hold the lever in position.

Further, the control rod from the dash lever could be inserted in alternate holes on the lever. One position, the long position, allowed all the above functions to occur. The short position prevented entering the float function.

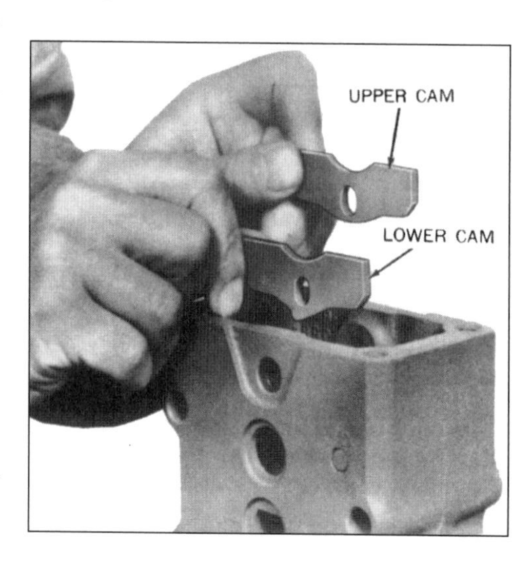

The selective control valve cams (shown being installed) are used for implement control functions.

Wilbur Davis, Chief Field Test Project Engineer, and Wendell Van Syoc, Field Test Engineer, related one 9-piston pump recurrent problem: "Without a reliable hydraulic pump the tractor could do nothing. The power steering, rockshaft, brakes, and remote cylinders were powered from the main hydraulic pump. The internal leakage also made the pump noisy. The hydraulic pump ran so hot that the red paint turned to a dull orange.

Ober Smith, Hydraulic Test Engineer, is credited with a pump heat test indicator. Temperature checks were made by spitting on the pump. If saliva sizzled, it was hot! If it erupted in a bubble of steam, it was time to shut down! Many times, the evaluation crew had to pull the tractor into the shop to remove the pump. There was more than one burn from the hot oil and pump."

Kenny Murphy, Field Test Technician, related an experience with a mounted corn picker on one of the first prototype tractors: "The excessive heat of the hydraulic system was rejected through the cooling system. Because of this we had to keep an extra 5 gallons of water on hand to cool the engine. The hydraulics had deteriorated so much that two persons had to help raise the picker gatherers."

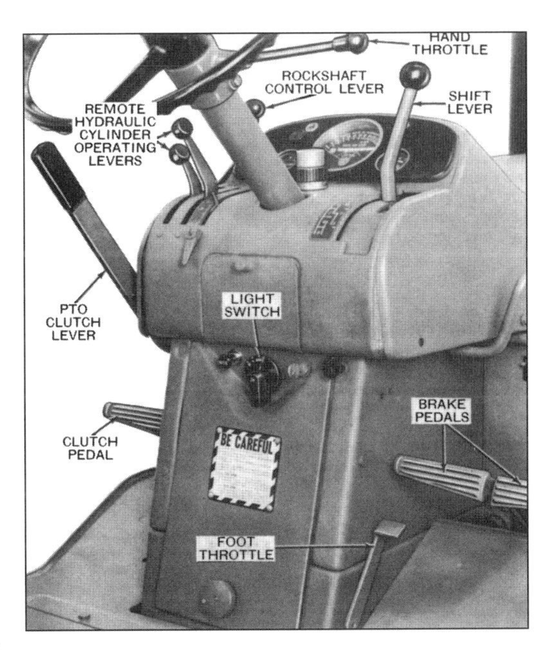

The levers and pedals are conveniently located for the operator on the 4010.

Some operators may use this feature to avoid an unexpected implement drop. Four poppet control valves were needed to control a double-acting remote cylinder. One inlet and one return valve allowed oil to flow to the head end of the remote cylinder on the raise cycle. Another pair allowed oil to return from the rod end of the cylinder. On the lower cycle, an alternate set of valves was actuated. The cams closed both inlet valves in the float operation thus connecting both ends of the remote cylinder directly to the reservoir.

Flow Control Valves

As development work progressed, it became evident that certain hydraulic functions were more important than others. With the safety of vehicle operation preserved, priorities were provided in periods of demand exceeding the pump flow. First priority was given to steering, second to the brakes, and everything else was last. Priority was controlled by appropriate sizing of the flow control valves in the function supply line.

The selective control metering shaft and outer valve regulated the amount of oil flow in slow raise or lower modes. A flow control valve was also used to regulate the oil flow in fast operation.

A flow control valve was also used to control oil flow entering the selective control poppet valves.

Main Rockshaft Control

The hydraulic valves used in the rockshaft were similar to those described for the selective control check valve and control valve. Principal differences were:

- The range of flow (6 to 7 gpm) made an adjustable orifice unnecessary. A fixed orifice through the valve center sufficed and adjustment was made by changing the spring force.

This view shows the selective control valve locations, operating levers, and links.

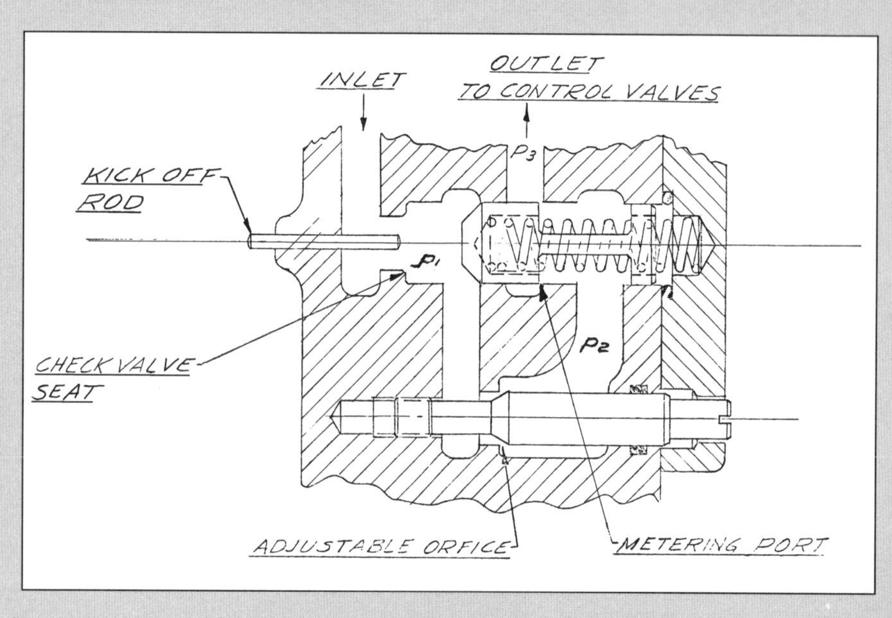

This drawing shows the inner working of a flow control valve.

Pressure drop across the adjustable orifice caused the movable spring-loaded valve to dither (oscillate) across the metering port cutoff. Movement of the valve to the right under pressure P1 was opposed by the lower pressure P2 and the spring load. If insufficient flow occurred, pressure P2 became closer to pressure P1, moving the valve to the left and allowing more oil to flow into the P3 space. Hence, this was a "pressure compensated" flow control valve. This meant that the flow rate was nearly constant regardless of the load that may cause pressure P3 to vary. (In

- A kick-off rod was unnecessary as there was no lever to return to neutral.
- A single pair of inlet and outlet valves were used as the rockshaft was operated by a single acting cylinder.

A principal new feature was the control means used to coordinate the mix of control lever position, draft signal, rockshaft position, and load and depth control. This again was a task that could be, and is, easily solved electronically today. Part of the solution involved a small bevel gear drive similar to a transmission differential drive. This mixing of signals was covered by patents that collected royalties in both the United States and Germany.

The cylindrical "bending bar" load control shaft was supported in special curved bushings in the transmission case sides and the drawbar support. The curved surfaces in the transmission case bushings acted to shorten the bending fulcrum point distance under load, thereby variably increasing the shaft stiffness. The load sensed by the bending bar was about 10% of the actual draft. The

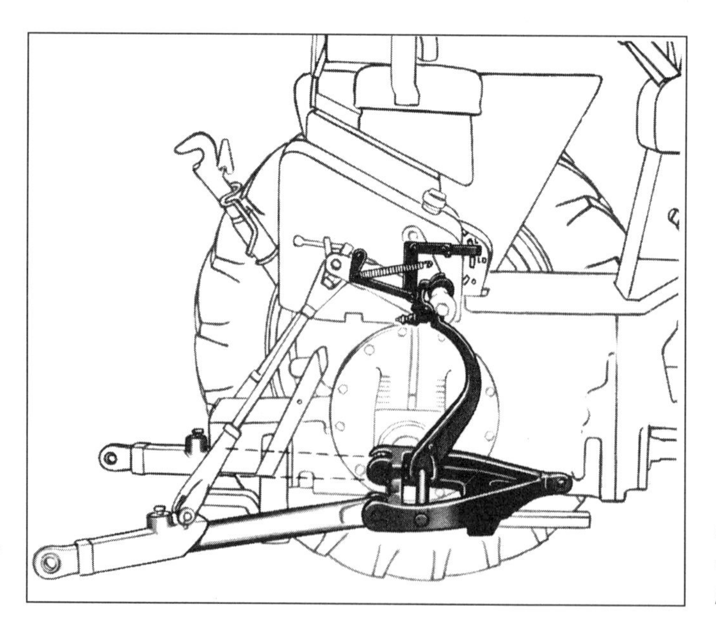

The draft links and sensing mechanism are highlighted on this phantom tractor view.

systems with pressure and volume compensation, the P2 pressure was also conducted to the main pump control to reduce the otherwise constant pump output pressure.)

When the remote cylinder reached the end of its stroke the pressure P3 quickly increased, sending the flow control valve to the left to its seat in the P1 chamber. It struck the kickoff rod, causing the control lever to return to neutral by lifting the detent cam holding the lever in fast operation.

movement of the bending bar in a negative draft direction was limited by a dowel pin stop in the bottom of the transmission case. Negative draft could occur when a heavy mounted implement was raised and under some unusual field conditions.

An oil seal was used where the bending bar passed through the transmission case wall. This lip-type oil seal was to retain the transmission and hydraulic fluid and exclude outside dirt, water, and mud. It was soon found that the seal required greater than usual flexing capability to follow the bending bar excursion. A small development program ensued to attain the necessary characteristics.

After several years in production it was found that the seat in the case for the seal outer diameter corroded. Customers were unhappy because the only provision for field service correction was to replace the entire expensive transmission case. Another design was soon provided to permit field reclaim of the seal seats at a much lower cost.

The outer ends of the bending bar supported an "A"-frame shaped casting to which the drawbar and hitch draft links were attached. The applications requiring drawbar draft control of implements was not as numerous as those requiring hitch draft link control.

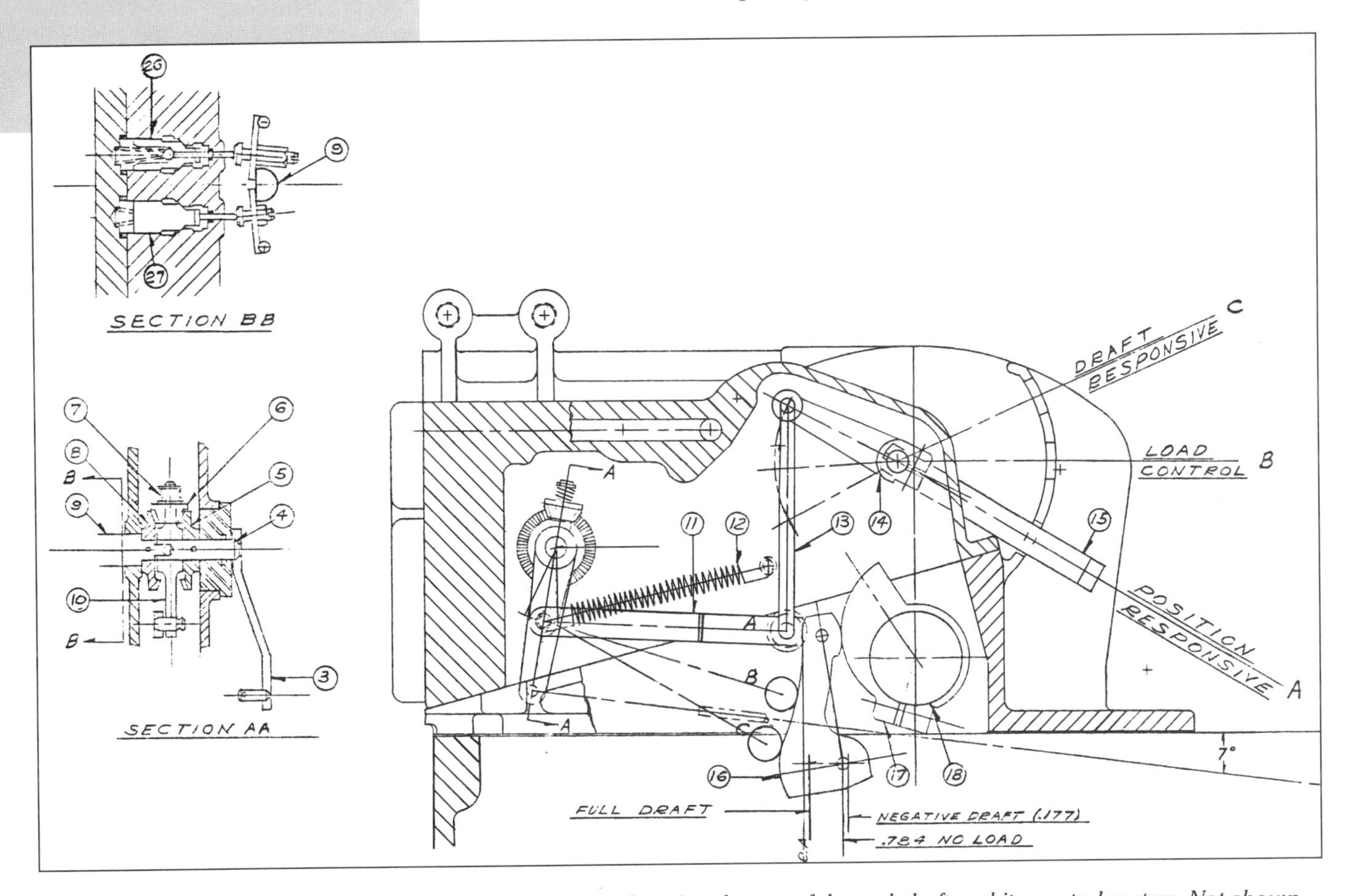

This illustration is a portion of a layout showing the key functional parts of the rockshaft and its control system. Not shown are the bending bar and lever that monitored the implement or drawbar draft load. The lower end of the "L"-shaped link number 16 moved in response to displacement of the bending bar under load. The upper end of link 16 contacted the rockshaft cam and signaled rockshaft position. The amount of draft signal being used was varied by the "load and depth" lever number 15, as shown by the respective A, B, and C positions. The combination of these three signals was transmitted by link 11 to lever 10, the upper end of which carried a small bevel gear 6. The control lever 1 through cable 2 and lever 3 controlled the rotational position of side bevel gear 5. The motions of gear 5 and gear 6 caused a resulting motion on gear 8. Gear 8 rotated shaft 9 which cammed the respective poppet valve 26 or 27.

The requirement for a load and depth control lever was determined in hitch development on the earlier two-cylinder tractors. Instances occurred where either all draft response, or no draft response, was desired. Most implement and soil conditions favored a mix of the two. Although infinitely variable adjustment of the load and depth lever was possible, a detented position giving about 60% draft response was available at lever position LD. (Depth control was sometimes called position control since the rockshaft control lever's position was the only control signal.)

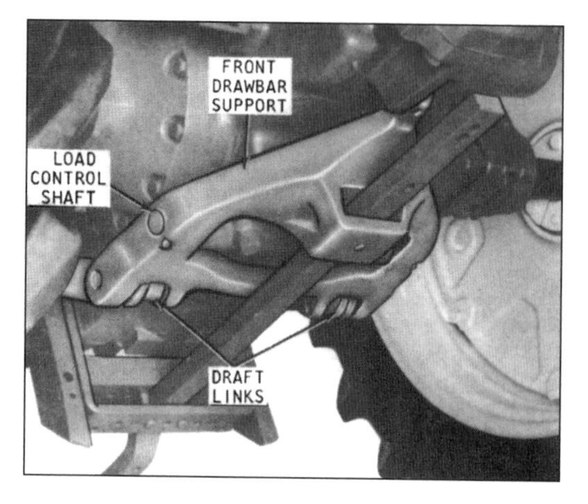

This underside view shows the tractor drawbar support and the load control shaft location.

Problems Arising in Test and Development

Loss of power and heating of the hydraulic system was possible through valve leakage in the closed center hydraulic system. Concentricity and finish of the valve seats were found highly important. Further, if not manufactured with sufficient precision these parts could not be reclaimed.

Another cause of leaky valves was sticking of the metering shafts (which lift the ball check valves). This problem was caused by corrosion induced by the chlorine and sulphur compounds used at one time in the extreme pressure hydraulic oil. Changing to alloy cast iron for the hydraulic pump pistons eliminated the need for the extreme pressure oil, and eliminated the corrosion problem.

Leakage also occurred if the valve castings were insufficiently cleaned. Wire or sand from the casting cores could break loose and damage valve seats. Attention to coring practice and design, and the use of Kolene cleaning process practically eliminated this problem. The cleaning process removed the core sand, loose dirt, and other foreign material from castings.

Proper valve adjustment, both rockshaft and selective control, was necessary for proper operation. This adjustment was required not only for initial operation, but also to correct for wear. Adequate and practical adjustment means and procedures were later developed.

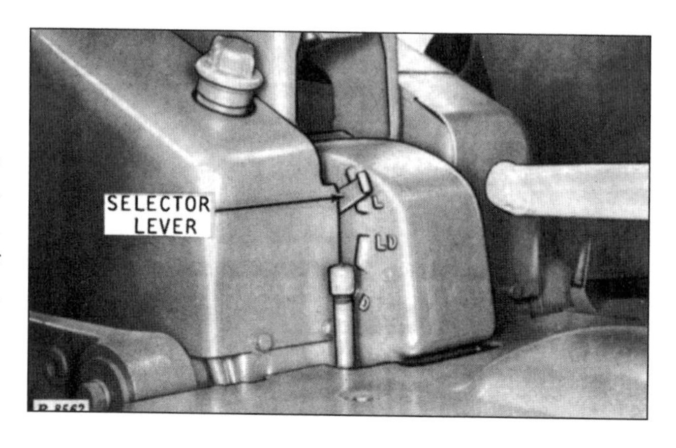

The need for a load and depth lever was determined during the hitch development of the earlier two-cylinder tractors. This illustration shows its location.

One of the first experimental tractors available for evaluation at the Laredo, Texas test site had all the prospective hydraulic functions. Besides the rockshaft, the fore-and-aft and lateral leveling features were included. These functions were controlled from the dash-mounted levers through automotive quality single wire push-pull cables. The unseating of closed-center valves required a certain amount of force.

These cables were inadequate, especially in the push direction. As a result they would buckle, then break after several cycles. Wendell Van Syoc related the attempted quick field shop repair: "Wilbur Davis and Harold Kienzle were attempting to repair the broken cable with an acetylene torch. After unsuccessful trials, they turned up the weld heat that produced a blob of molten steel on the wire end. This also produced some colorful language from Wilbur."

Later, multiple strand cables were successfully used. While the cables provided adequate strength, some development steps were necessary to adapt the housing-to-cable seal to the tractor operating environment.

These new tractors used hydraulic tubing connecting the various hydraulic components to a degree not previously used on the two-cylinder tractors. Further, the higher operating pressure created learning experiences in tubing inspections, fastener design, and tubing fatigue. Wendell Van Syoc related an experience that led to a tube fastening improvement. "One morning in Laredo, Merlin Hansen was first out to drive an OX tractor. As the engine started, a hydraulic line connection on the rockshaft split open under the seat. Oil ricocheted all over from fenders and cowl. An oil-soaked chief engineer went to the shop and placed a call to Harlan Jensen in Waterloo. We soon had double flare joints that solved the problem."

The John Deere Model 720 two-cylinder tractor had the concealed steering motor.

Power Steering

The power steering system was another departure from the system used on the two-cylinder tractors. For years (from the 1933 Model A), the two-cylinder row-crop tractors could be characterized by the almost horizontal steering shaft. It had an almost vertical steering wheel on the rear. The steering shaft entered on the front a cast-iron housing for worm and sector gears. This housing was placed atop the front casting attached to the frame and front wheels. In the early 1950s power steering became available. It was so well blended into the structure under the styling sheet steel that it was not perceivable. In 1958, the 30 series two-cylinder tractors were introduced with a more horizontal steering wheel. A universal joint at the rear of the then-enclosed steering shaft aided the steering wheel.

The new tractors placed the fuel tank atop the steering spindle. Because of the need for engine service access, the top placement of the steering shaft was eliminated. Possible system arrangements included: drag link servo cylinder system, flexible cables, chain and pulleys, complex gear and shaft arrangement, and a hydraulic system.

Drag links were not possible in the design as there were to be no projections outside the sheet metal or frame. The other arrangements, except the hydraulic system, were believed to have backlash wear, or unacceptable looseness for positive firm steering, as well as being difficult to maintain.

The all-hydraulic system without any mechanical connection between the steering wheels and front wheels was the agreed choice. The flexibility for any tractor arrangement was obvious. Getting a satisfactory system took several design changes.

The John Deere two-cylinder Model 730 tractor had the angled steering wheel.

Ed Fletcher, Hydraulic Project Engineer, when reflecting about making the all-hydraulic decision, said: "We did lots of driver testing and I lost lots of sleep over the 'no mechanical' system. Much effort was given to keeping the goal on driver feel."

A spool valve to control the oil for steering was initially tried. Selectively assembled valve clearances of 0.0004 inch were necessary to control leakage. Then the valve function could be impaired because of housing distortion from mounting, temperature changes, or dirt in the oil.

A spool valve and gear pump mounted on the opposite end of the steering wheel shaft was also tried. It steered the tractor but had problems with leakage causing lack of index between the steering and front wheels. In addition, both the valve and gear pump required extensive sealing to be compatible with the 2,000-psi hydraulic system and to cope with wear.

Better leakage control and creeping was obtained when a new Racine vane pump and motor was used. The final system used a master and slave cylinder hydrostatic system. The master cylinder, on the bottom end of the steering shaft and wheel assembly, was moved by rotating the steering wheel through a screw mechanism to open and close paired poppet valves. These poppet valves were similar to those used on the selective control valves. Pressure oil was admitted to one side of the double-acting master cylinder. Oil in the other side was sent by tubing to the steering motor in the front of the tractor. Turns in the opposite direction operated another set of poppet valves and applied pressure oil to the opposite master cylinder side.

The steering motor consisted of two double-acting pistons (in the slave cylinders) that in the center of their length had rack teeth. These teeth engaged a pinion which was part of the front steering spindle. The volume of the two slave motor cylinders was exactly equal to the master cylinder volume. The piston and gear mechanism was housed in a large casting to which the front end of the side frames were attached. This casting also provided the support for the front wheels on tricycle tractors. The lower end of the steering spindle on wide front axle tractors had connections to the respective steering tie rods.

Operating characteristics of the gear pump type power steering system have been related by Wilbur Davis: "This system had the disconcerting habit of unpredictably 'locking up.' This meant that the gears were held against the side of the pump housing and could not be turned with the steering wheel. Before one of the early demonstrations, road testing the latest design was prudent. I (Davis) drove the tractor at progressively higher speeds, ending in eighth gear, much to the concern of observers."

Leakage was another problem inherent with the gear pump concept. According to Wendell Van Syoc: "The steering wheel position drifted relative to the front wheels. It was necessary to continue making steering wheel position corrections to keep the tractor on a straight course."

This phantom 4010 tractor view highlights the steering valve, motor, and lines.

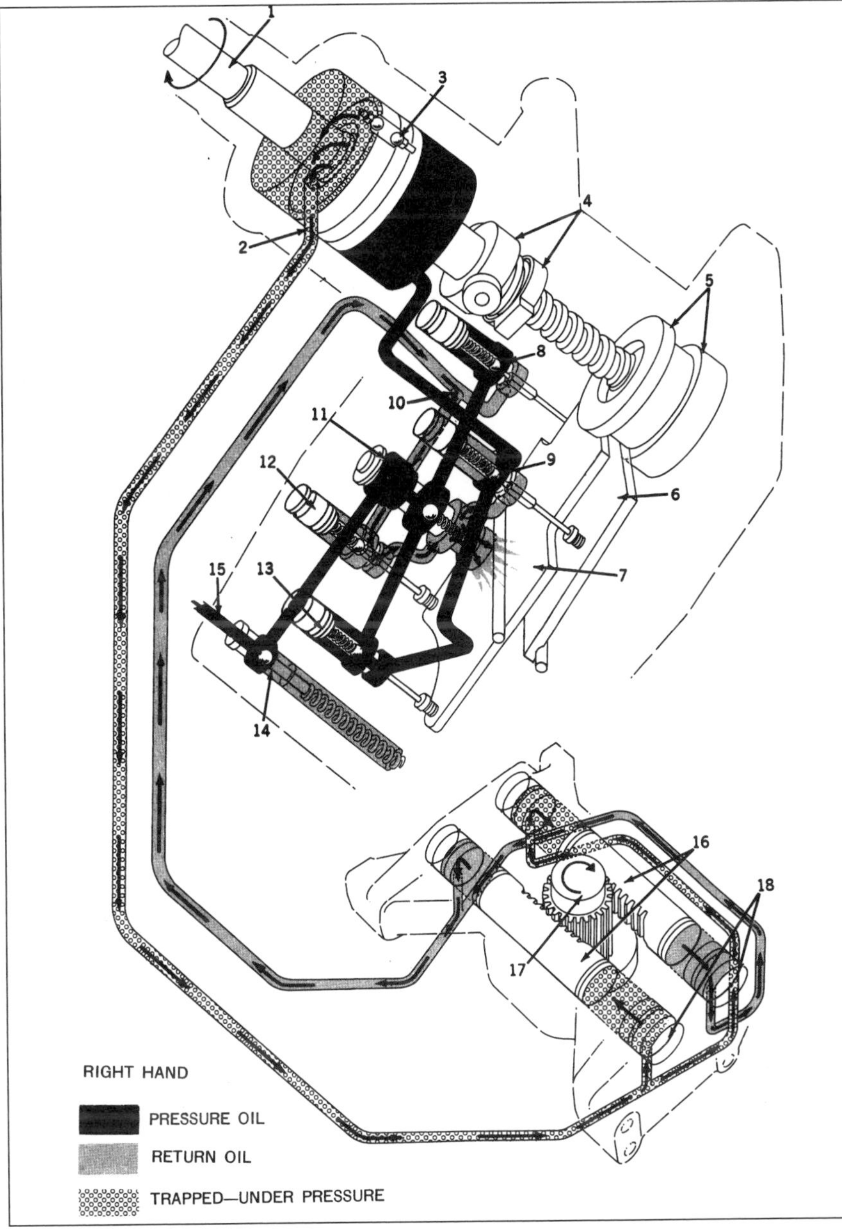

The steering system in a powered right hand turn is shown in this drawing.

A provision was included for manual steering if a hydraulic system failure occurred. The screw drive, previously described, could move the master cylinder piston. An unloading valve disconnected the steering system so the oil used for manual steering could not flow back into the tractor hydraulic system. A pressure relief valve to protect the system parts was also included. Manual steering caused more steering effort and was not intended for long-term use, but did allow the operator to steer the tractor to a stop.

One other provision was included to correct the steering indexing between the steering and front wheels. One of the many O-ring seals could wear or become damaged in assembly or by system debris. Oil in one side of the master-slave system could be lost and insufficient steering could result. A provision was included such that with a full turn to one side the captive oil system would refill. When this occurred the front wheels made a rapid steering motion at the end of the turn.

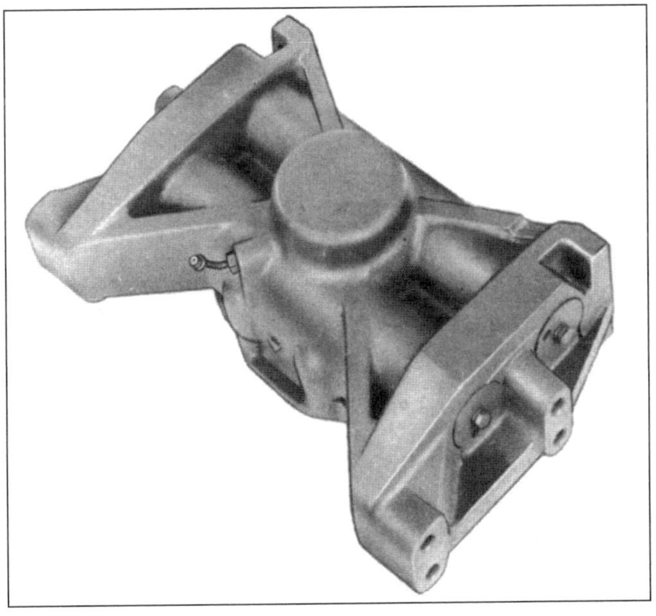

The steering motor consisted of two double-acting pistons that had rack teeth.

Brakes

The new brake design used concepts completely different from the two-cylinder tractor. Ideally, brakes are on the differential shafts so the torque requirement is less than on the final drive or axles. This arrangement makes the brake package smaller than if the brakes were mounted on the axle shaft. Of course, the two-cylinder tractor brakes were not an ideal objective. Not only were they manually operated, they often had poor durability if braked turns were frequent. The linings had to be frequently replaced. On a new tractor the paint on the formed steel brake drums was soon scorched brown. A longer life approach was needed. Commensurate with the progression to larger tractors, power application of the brakes was desired. The chosen brake location was mounted on the final drive sun pinion driven by the differential, before the final reduction. This was the equivalent of the reduced torque mounting of the two-cylinder tractor brakes.

The power brake system is highlighted in this phantom tractor view.

The brake pads that applied pressure on the brake disks were intregally attached to the brake pistons. The brake pistons in turn were installed in machined recesses in the transmission case final drive compartments. The brake pistons had O-ring seals on their outer diameter. The slight rolling action of the O-ring, when under pressure, returned the piston when the pressure dropped. This action provided sufficient retraction of the brake piston to eliminate drag when the brakes were not being used. The oil passages for the brake piston supply were drilled in the transmission case. In summary, this was a very compact, efficient design with high capacity. Thus, the brake system was another of the innovative concepts that enabled the total package to meet the stringent program objectives.

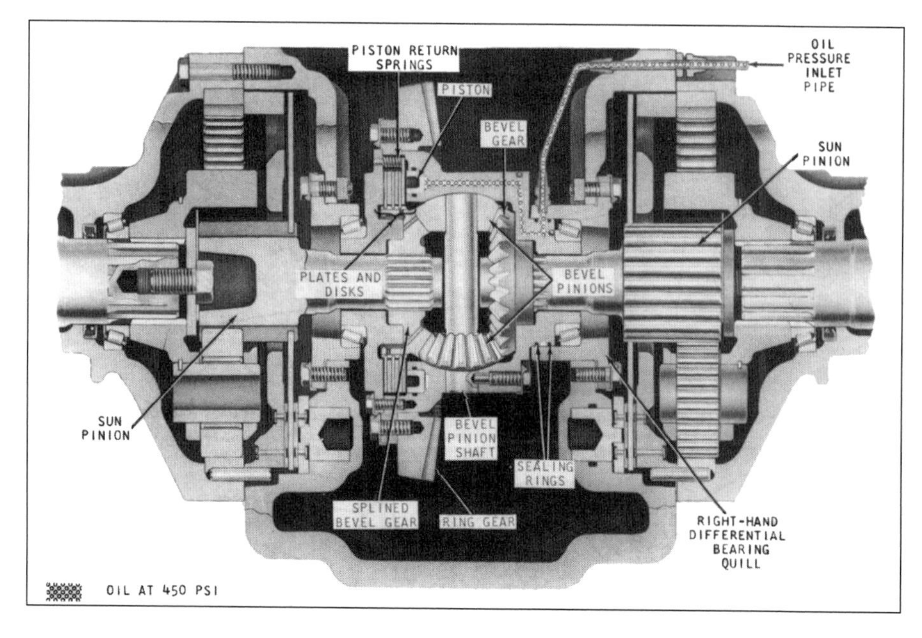

This cutaway drawing shows the final drive and differential assembly with the differential lock.

The requirement for longer brake life was met by the brake pads operating in oil for cooling. Considerable testing of brake pad material was needed. A material hard enough to give good wear exhibited a "stick-slip" chatter characteristic. Surface finish of the brake disks was important for low wear and chatter reduction. The brake chatter was further resolved with an oil

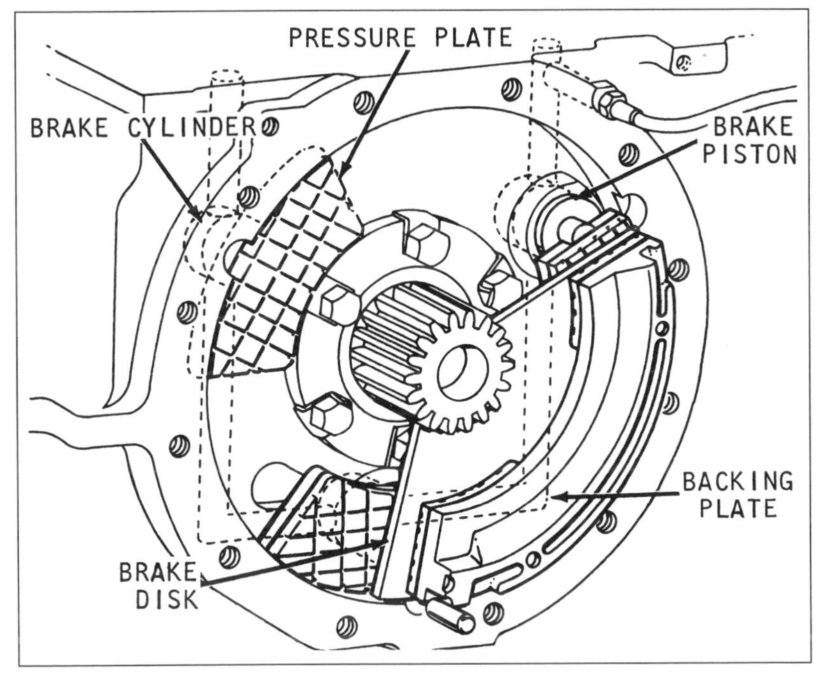

This phantom view illustrates the brake parts.

additive "wetting agent" which was a major ingredient of the special hydraulic oil.

The brake pistons were operated by oil coming from the master brake valve. A valve, piston, and connecting brake pipe was in place for each side of the tractor. This poppet-type valve provided load-proportioned feedback to the operator's foot through the respective operating pedal. A latch between the pedals allowed the brakes to act in unison for safer highway travel.

Two further provisions were incorporated. One was for operating as a manual brake should the main hydraulic system fail. As may be expected, higher pedal forces were required for manual braking. A maximum of two strokes was the objective to get manual braking activated. The other provision was a procedure to bleed air from the system to eliminate a possible spongy feel and safety problem.

This view shows the brake pressure plates.

Edward Fletcher, Chief Project Engineer for Hydraulic and Hitch Design (*Product Engineering Seminar 1959*), emphasized the need for education of the operator, customer, implement factory, and service persons to gain acceptance of the new hydraulic system:

> *"We should promote the use of cylinders and motors to do various little jobs where it is more convenient. This is because they have little affect on the other functions of overall efficiency of our constant pressure system. We should encourage the use of smaller cylinders and motors to keep operating pressures near full load to avoid inefficiency under light load conditions. We should encourage the efficient use of metering action to start and stop loads. This will give smoother operation and avoid shock loads on the tractor, implement, and hydraulic system."*

The hydraulic system, of all tractor functions, had more innovative concepts. Scattered all over the tractor, each component is limited in space, and considerable ingenuity is evident. In the author's opinion, the hydraulic system, and its components, helped most with enabling that the tractor design objectives were met.

This close-up shows the brake disk mounted on the sun pinion shaft.

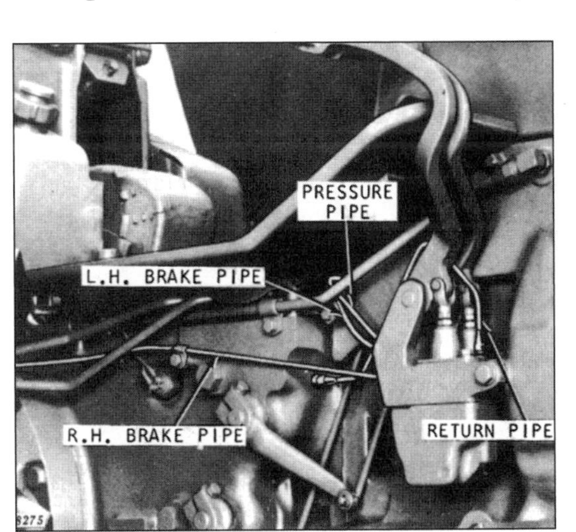

The external pipes of the brake system are shown in this close-up shot.

Implement Hitch

Mounting farm equipment on tractors has been popular since the introduction of the International Harvester Farmall in the 1920s. Until 1934 the attached equipment lift was obtained by mechanical or manual means. That year, however, John Deere introduced the model A tractor with the hydraulically-powered implement lift called the Power Trol. This lift system, principally designed by Emil Jirsa, raised tractor-mounted implements that functioned integrally with the tractor. The implement depth adjustment by the operator had no feedback system to adjust for draft. Attachable lift arms and "push-pull" tubes were used to lift, lower, and adjust the working depth of front-mounted implements.

In 1935 Harry Ferguson introduced a load compensating implement hitch on Ford tractors which used three powered links having attaching points for rear-mounted implements. This hitch differed from the competition in that it:

- provided a tractor mounting of normally-drawn implements such as plows and disks;
- provided an automatic means for varying implement depth according to draft load on the tractor; and
- enabled lightweight tractors to increase their draft pull by dynamically transferring some of the front end and implement weight to the rear traction wheels.

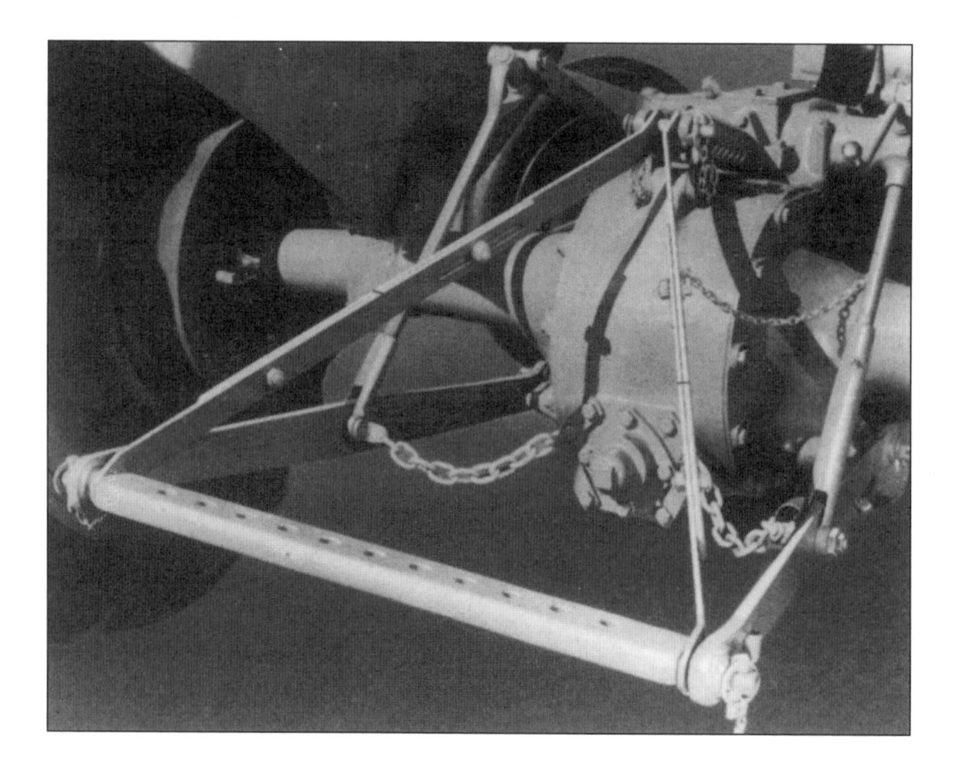

In 1935 Harry Ferguson introduced a load compensating implement hitch for Ford tractors which used three links with attaching points for rear-mounted implements. The Ferguson System hitch is shown here on the Ford 9N tractor.

During the 1950s Ford developed larger tractors featuring the new hitch. An even larger hitch was available on the British-built Fordson Major diesel tractor.

This system also offered lower cost implements, improved ease of attachment, provided easier transport, and reduced the tractor's added ballast requirements. The hitch action was somewhat harsh by comparison to today's designs and the operator was aware of each adjustment by sound and seat forces. At times the system tended to be over responsive and hence caused unnecessary variation in implement depth. In field demonstrations these units were quite impressive when compared with older and heavier tractors using towed implements. These advances helped the continuation to lower weight and higher horsepower-to-weight ratios in tractors.

Customer interest was slow at first, but as popularity increased, all competitive tractors introduced over the following years provided comparable hitch designs. Ferguson and Ford divided into separate companies in the late 1940s. In the 1950s Ford developed larger tractors, all with the hitch. About 1953 Ford made a larger hitch available on their British-built Fordson Major tractor. With wider attachment points, its dimensions were similar to those of the later Category II standard.

In the 1950s Deere & Company (as well as other competitors) began developing hitch-mounted larger implements for the existing larger tractors. An intensive program, involving both the tractor and implement factories, tested several candidates for production use on the two-cylinder engine tractors. Deere recognized the need for heavier hitch members, greater width between rear draft link attaching points, and a greater mast height. Activity was initiated by Deere to develop a FIEI (Farm Industrial Equipment Institute) standard for a Category II hitch which was convertible to Category I. This standard enables tractors of any make to attach conveniently to implements of all makes. The hitch dimensions are standardized by the American Society of Agricultural Engineers (ASAE) and the Society of Automotive Engineers (SAE). Tractors are typically used with a variety of different makes, as well as types, of equipment. Industry agreement on the hitch connections expedited the furthering of hitch use. The ASAE-SAE standards helped accomplish those goals (see ASAE Standard 217 first adopted in March 1959). Category I and II hitch standards existed during the OX (4010) and OY (3010) design program. Larger Categories III and IV standards have since been approved.

The first hitch Deere introduced into production was a draft responsive Category I hitch for the Dubuque-made 40 series tractors in the early 1950s. The first hitch offered in production on the larger tractors was the 800 hitch made by the John Deere Plow Works. This hitch provided the desired ease of attachment and transport. The gauge wheels controlling the implement depth altered the line of draft and amount of weight transfer. It was introduced in 1953 on the 50, 60, and 70 tractors. A later version, the 801, gave some greater mechanical load response, but was only built and warehoused. It was never sold since it

This Model 40 John Deere tractor has the first Deere draft responsive hitch connected to a mounted disk harrow.

became available at the time of introduction of the Waterloo-built weight transfer hitch. This latter hitch provided a mechanical linkage conveying a draft signal to the lift system.

The hydraulical draft responsive hitch, called the Load and Depth Control hitch, was made available in 1956 on the 20 series two-cylinder tractors. This hitch provided a selection of either precise operator control of implement depth or height, or automatic adjustment of working depth for uniform tractor motion. There was also an intermediate position that gave a mix of both. This hitch, the first for large John Deere tractors, sensed the implement load and automatically corrected with a depth change. The same hitch continued to be available in the 30 series two-cylinder tractors.

The design of the OX and OY tractor hitches was different from the 730 tractor hitch in several respects. These differences included: a load signalling system, a draft link construction, sway control means, lift arm to lift link connection, and a new implement attachment means — the quick coupler. Other new components in the OX and OY equipment attachment realm were new front frame weights, a front rockshaft, and new implement attachment locations.

Load Signaling and Control Means

The hitch on the 20 and 30 series two-cylinder tractors (and the Ford Ferguson tractors) received their load signal by monitoring the forces in the top link. The usual three-point hitch mounted working implement produced a rollover force about the lower link attachment. The results were tension forces in the lower links and usually compression forces in the upper or center link. These compressive and tensile forces and their motion were resisted by a large helical coil spring. The force in the spring signalled the hydraulic depth control mechanism.

There are load conditions when the upper link acts in tension rather than compression. Roy Morling has a theoretical explanation of this effect entitled *Agricultural Tractor Hitches: Analysis of Design Requirements* in ASAE Distinguished Lecture Series Tractor Design, No. 5, 1979.

With lower link signalling, a means of averaging the force difference in the two links was required. To accomplish this task, the OX and OY designers chose a bending bar with fulcrums in the transmission case sides. The bar's outer ends were in the front drawbar support casting. The drawbar support in turn had attaching points for the front ends of the two lower links. The signal was taken from the center of the bar, which averaged the signal in the two lower links.

The first design attached the draft links directly to the outer ends of the bending bar. The front of the draft links could be shifted in or out to change the Category. This design had the advantages of using a common sway block setting for either hitch Category. It also automatically provided increased sensitivity for smaller Category I implements. The concept was dropped, however, due to the difficulty and inconvenience of getting under the tractor to change Category settings.

Also, mounting the drawbar support on the bending bar made possible a draft responsive drawbar. This feature was desirable when the tractor was towing implements whose working depth was controlled with hydraulic remote cylinders. Testing the draft responsive drawbar did not provide the expected improvement in field performance.

The bending bar fulcrum supports in the transmission case and drawbar support were hardened steel bushings that were specially shaped on the inner diameter. The shape moved the points of support in the drawbar support casting inward as the load increased. This movement reduced the effective bending length of the bar. It gave the signal a variable sensitivity and eliminated the need for a separate sensitivity control as used on the 30 series hitches.

The bending bar ends passed through the sides of the transmission case below the transmission operating fluid level. A specially designed seal, using a garter spring, with a bellow "S"-like construction, followed the excursion of the shaft.

When lifting heavy implements or under unusual soil conditions, a negative draft signal sometimes occurred. This signal could inhibit its ability to raise the implement. Overstressing the bending bar by extreme negative signals was prevented by a dowel pin stop in the transmission case.

Also, a stop in the drawbar support attachment limited the bending bar excursion and potential overstressing under load in the positive draft direction.

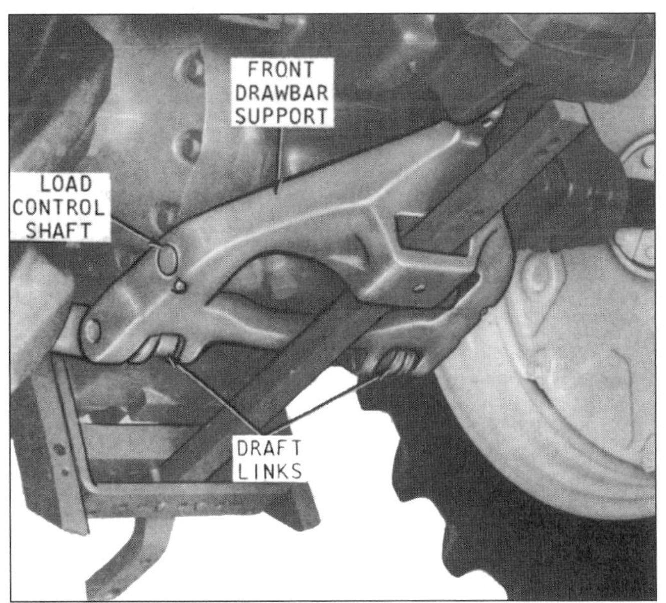

This illustration shows the drawbar location in the front drawbar support. To accomplish the lower link signalling, the bending bar with the fulcrums in the transmission case was required. The bar's outer ends were in the front drawbar support casting.

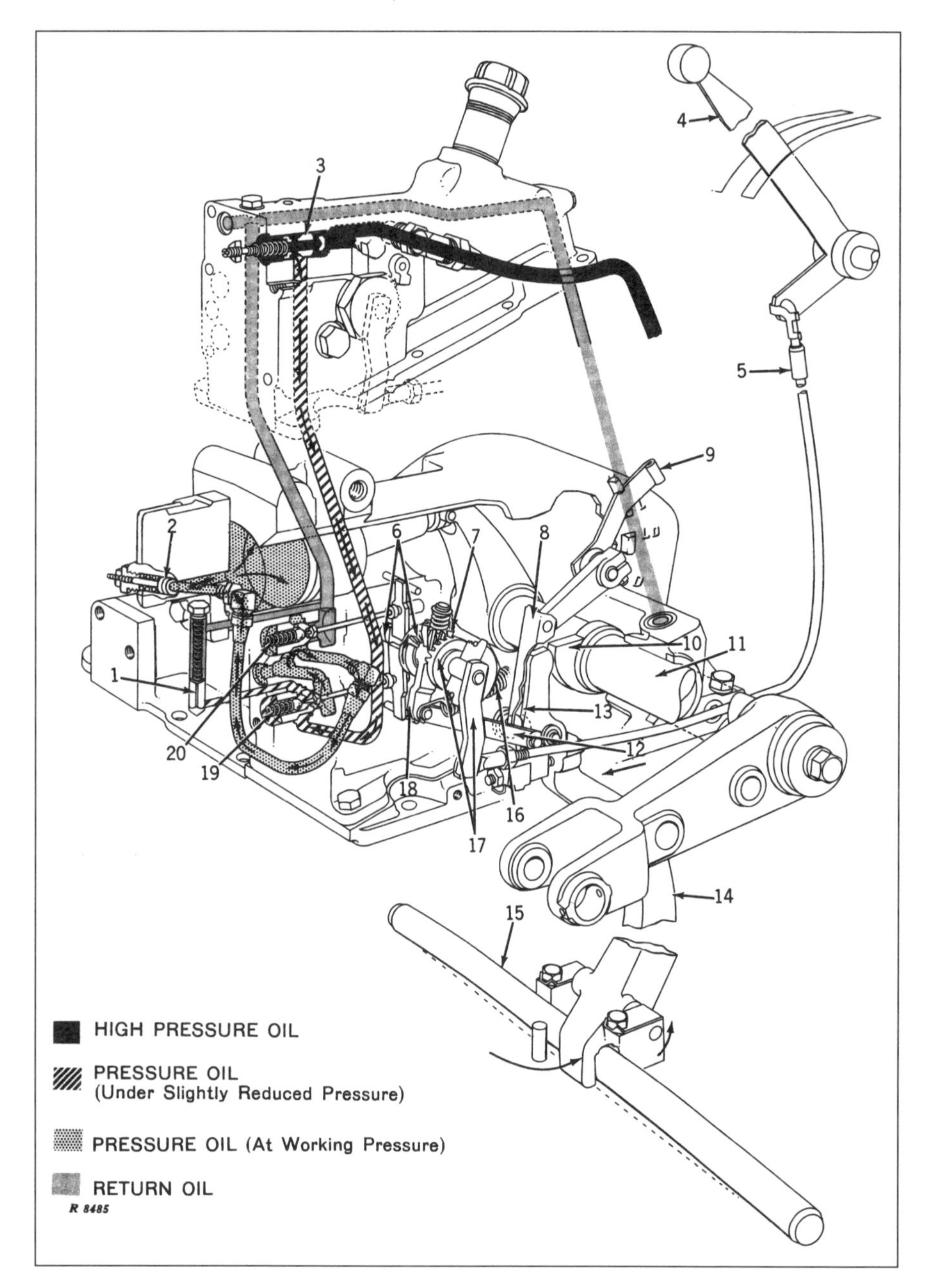

This phantom view shows the rockshaft assembly with the oil system highlighted. The negative signal dowel pin stop is also shown.

Hitch Geometry

Hitches were more than just a means of lifting and carrying an implement. The links and their attachment locations permitted the implement to follow the tractor over small hills and through depressions or swales. The geometry of the hitch to do this in the vertical plane was similar to that of a parallel linkage. Designing the hitch geometry involved a compromise considering several important characteristics, which included lift capacity, hitch sensing, transport clearance, implement penetration, hill-and-swale and contour operation.

The upper and lower links converged in the vertical plane toward the front part of the tractor. This convergence was necessary to tip the rear end of the implement up at the end of lift. It also enhanced the ability of long implements to maintain a uniform depth when operating over hill-and-swale and also had a significant effect on the ability of the implement to penetrate in hard ground.

In the horizontal plane the links enabled the implement to trail a tractor through gentle turns. This feature reduced the tendency for the implement (primarily moldboard plows) to overcut or undercut when operating on contoured land. The links accomplished this task by having a line of force convergence toward a point forward of their attachment. These points of convergence in the horizontal and vertical planes are not common. See the Morling reference (on pages 74 and 75) for a theoretical presentation of this theory.

The success of the hitch geometry defined the acceptance of the hitched implements in the field. The hitch geometry of the OX and OY tractors was very similar to that of the 730 tractor, which had been judged very acceptable in the field. New equal rolling radius tires were provided with all tire width options. The new constant radius tires permitted greater lifting capacity by allowing the implement always to be the same distance from the tractor.

Draft Link Construction

For lower cost the draft links on the 3010 (OY) tractor were very similar to those used on the 730 tractors. In addition to the solid draft links, a straight telescoping draft link was available. Both styles had been introduced some years earlier on the two-cylinder tractors. These telescoping links permitted limited fore and aft motion allowing easier connection of the ball end. The tractor could then be moved backwards, reducing the telescope distance until a spring-loaded pin dropped into place. Under this condition the draft link then functioned as a solid link. The solid links had a hollow square middle tube with welded steel ball sockets on the ends. For the telescoping version, a solid sliding square steel bar was inside the center of the square tube.

The OX and OY tractors also used a square bar. The square bar acted as a spreader to hold the lower links to the proper width. The bar had a center guide to position the lower links in the horizontal plane. After the lower links were connected, the center link with a "parrot beak" end was dropped over the implement mast upper pin.

Parked heavier implements, which tended to settle from the position where it had previously been disconnected, were more difficult to line up to the tractor lower or draft links. The tractor may also have been moved backward at an

The telescoping links, as shown in this three-point hitch, allowed limited fore and aft motion and therefore allowed for easier connecting to the implement ball end.

These two photos show the sway blocks, heavy duty telescoping draft links, and square spreader bar. The top photo shows the sway blocks in the down position to prevent side sway of the implement. The bottom photo shows the sway blocks in the up position allowing the trailing implement to sway relative to the tractor.

angle to the previous disconnecting lineup, which made it more difficult to line up. On small equipment the implement could usually be manually moved to enable attaching the lower links.

A more difficult connection of heavier implements for the OX tractor was expected. The objective of the new heavy-duty telescoping draft links was to make this task easier. These links allowed some vertical movement, as well as fore and aft movement, to give greater assistance in attaching to implements.

Sway Control

Some equipment could not use the sideways sway allowed by the draft links. As a result, some means were necessary to limit this sway. Earlier tractors, both Deere and competitors, used chains or pivoting links to control sway.

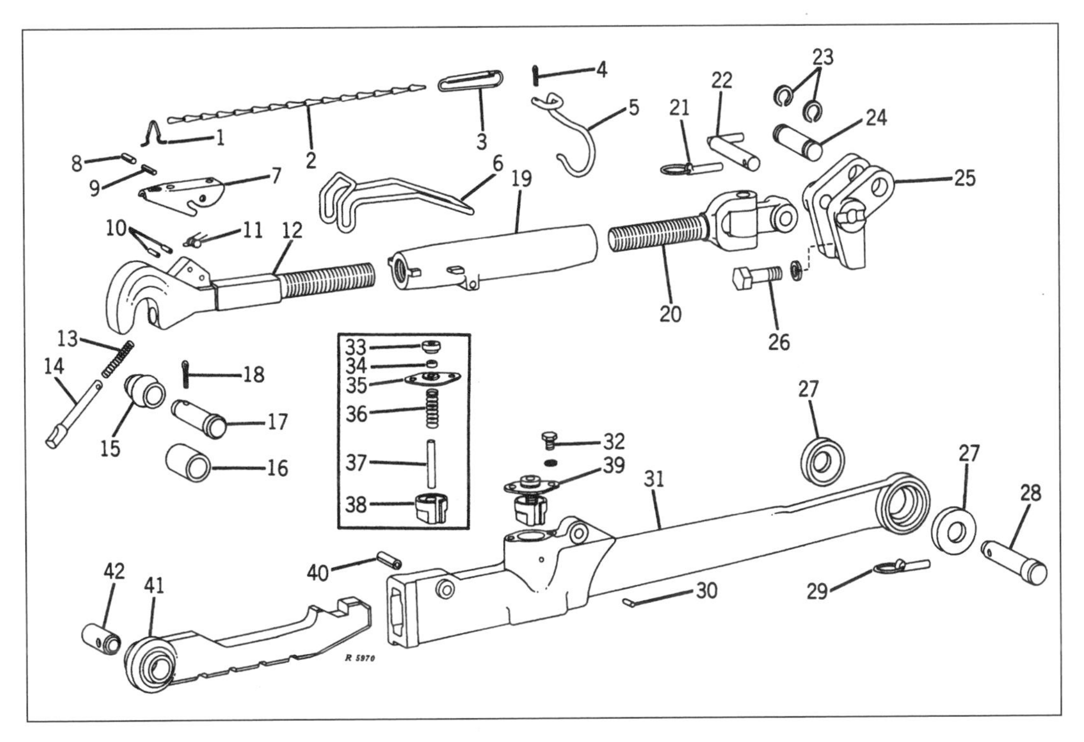

This exploded view shows the center link and telescoping draft link.

As explained earlier, at the time of the OX and OY design, two different sized hitch categories existed. The smaller Category I was similar to that represented by the original Ferguson hitch. As tractor powers increased, larger implement attaching pins became necessary. Also a wider horizontal distance between the hitch balls was desirable for implement stability both in the field and in transport. Sway lockout was required in transporting mounted implements. Hence, four conditions of sway were necessary for these tractors that worked with both Category I and II implements.

New sway blocks were designed to take care of these problems. The attaching bolts for all four positions were the same to avoid them being misplaced. For some operations where no sway could be tolerated, sway block shims were available.

Lift Link Float

Some wide implements used gage wheels to control the depth of the working tool. With fixed lift links these implements sometimes caused high torsional forces in the rockshaft and excessive lift link compressive forces. Avoiding this could be done by optionally allowing some loose motion in the lift links. An adjustable collar and pin allowed 1 inch of lift link float or lockout.

The adjustable collar and pin allowed 1 inch of lift link float.

Quick Coupler

The quick coupler was an innovation designed to lessen the operator's trips between the dash controls and the hitch ball point attachments. The quick coupler's goal was to enable the operator to connect to hitch-mounted implements without leaving the seat.

The chosen construction was an inverted U-shaped frame. The upper mid-point had a hook that picked up a cross shaft near the upper end of an implement's attaching mast. The lower ends had larger hooks to receive the implement's cross shafts near the lower link ends. The lower hooks had a spring-loaded latching plate controlled by handles reachable from the operator's seat with the hitch raised.

The principle of operation was to maneuver the tractor and hitch to engage the upper hook, then lift the hitch and implement. This movement caused the implement to swing forward to engage the lower hooks into the latching position.

The coupler did not change the three-point hitch configuration and motions. The implement required minor changes to accept the coupler hooks. The implement was moved further to the rear of the tractor by several inches. This relocation did cause heavier front tractor weights with some already heavy implements. However, it was popular with the field test personnel who often changed implements. And its popularity continued with customers such that a hitch coupler has since been available on all models of Waterloo-built tractors.

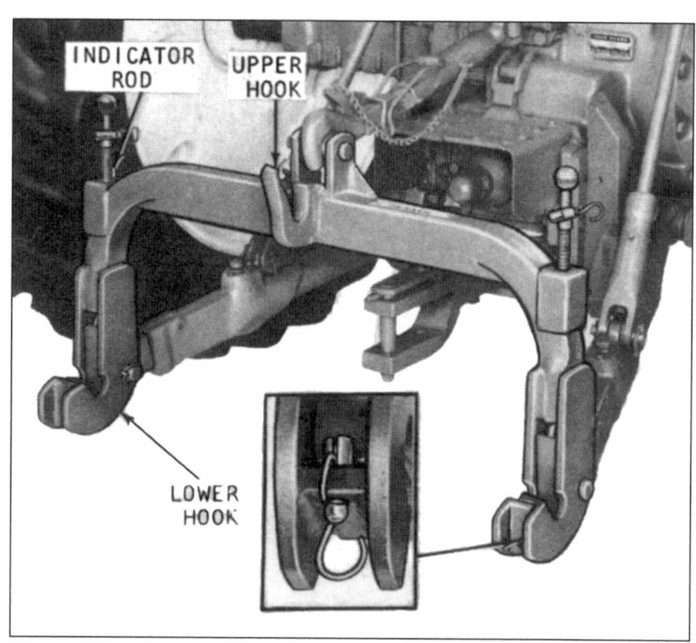

The quick coupler's goal was to enable the operator to connect to hitch-mounted implements without leaving the seat. The three-point hitch and quick coupler are pictured here.

Both Category I and Category II couplers were provided. Later a convertible coupler was adapted to both implement categories.

The configuration of quick couplers, as defined by ASAE Standard S278 first adopted in December 1964, promoted an interchangeable attachment with other manufacturer's implements. International Harvester gave considerable pressure to have their Quick Hitch adopted as the standard. Instead, the Deere Quick Coupler design was adopted and was believed to be adaptable to a wider array of implements.

Drawbar

The standard swinging drawbar had an offset bend of 1 3/4 inches to meet the ASAE-SAE (see ASAE Standard S203) standard drawbar to PTO dimension. The drawbar also could be used in the up-turned position that gave slightly more weight transfer to the rear wheels.

As explained earlier, the front of the drawbar was attached to the front drawbar support. The rear of the drawbar was supported by a U-shaped hardened steel support bolted on the rear of the transmission case.

Some implements, usually used in wheatland operations, required a roller-supported wide swing drawbar. This optional drawbar prevented use of the three-point hitch because of the wider drawbar excursion.

Hitch Lifted Drawbar

Considerable research was done using the hitch to lift heavy implement tongues for connecting to the tractor drawbar with a ball and socket attachment. The goal was to enable the operator to connect and disconnect this type of implement without leaving the seat. Despite the obvious advantage, no successful design concept arose that was acceptable to both tractor and implement designers. The design of the ball and socket was extremely critical in providing the necessary clearance for articulation and strength for towed implements. This study continued for about 10 years before being dropped.

Front weights and frame weights were required when using hitch-mounted implements.

Front Weights

Hitch-mounted implement acceptance required the addition of removable weights to the front of the tractor. The amount of weight added was a consideration when selecting front tires with adequate load capacity. To meet the added weight requirement, weights of any shape or style would work. However, meeting acceptable clearance with all possible implements placed limitations on the space and shape of the weights. Within those limitations, consultation by Henry Dreyfuss Associates made the appearance pleasing. Each removable side or front weight weighed 85 pounds each. Human factors suggested that lifting 85 pounds was the maximum desirable weight.

Further, the design configuration could be used on all the Waterloo and Dubuque (4010, 3010, 2010, 1010) built tractors. The front weights were also used on the more recent larger Waterloo tractors, sometimes with a revised side starter weight. This feature reduced the number of different weights needed in inventory by dealers and customers.

Leveling Control

One feature designed and tested extensively, yet did not go into production, was the automatic power fore and aft leveling control. The feature was demonstrated to Deere & Company management in 1953 at Brenham, Texas, by Harold Sexton and Wilbur Davis.

This design required a second rockshaft and piston, and an automatic control system. The goal, which it met, was to enable long implements to follow a sharply changing ground surface profile. The feature would have been costly, however, and the expected market volume would have been too small to justify the investment.

Front Rockshaft

For many years, raising and lowering front-mounted implements such as cultivators, used the rockshaft on the rear of the tractor. This movement required using push-pull shafts from the implement to special connecting locations on the rear rockshaft arms. These push-pull tubes interfered with access to the operator seat from the ground, and limited the use of some rear attachments.

The two-cylinder tractors had a front rockshaft option. It eliminated the need for the push-pull tubes and simplified the system for some front-mounted implements. A front rockshaft option continued for the OX and OY tractors. Remote cylinders were installed on the tractor with the rear of the cylinders attached to rear brackets on the tractor frame. The rod ends of the cylinders were attached to the rear set of holes in the front rockshaft arm. The remote cylinders on each side were coordinated by the rockshaft, which could be disconnected for independent side control. Control of the remote cylinders was by the dash-mounted control levers. (Note that the location and use of this front rockshaft is different from that used on later European-style front hitches.)

Older implements requiring the push-pull tubes could still be used with the new tractors by an available rear rockshaft arm.

Gradually the implements began mounting the remote cylinders entirely within their structure and so this method of attachment declined in popularity.

The front rockshaft eliminated the need for the push-pull tubes that had been used on the earlier tractors and simplified the system for front-mounted implements.

Implement Attachment

Some may wonder how various implements made in Deere factories became attachable to tractors made (now) in several other factories.

A "Tractor and Implement Committee" existed long before the start of the OX and OY tractor designs. The members were appointed representatives of the various factories. They were assigned to coordinate the attachment provisions and requirements, and also to provide the medium of exchange of information on future tractor and implement changes.

The OX and OY tractors offered a quantity of attaching points for implement usage. Fortunately, the dimensions carried over from the previous tractors were the frame widths, side frames bolting patterns, front attachments, three-point hitch, drawbar, and PTO. These similarities allowed the owner the option to use his older implements with his new tractor. These points provided three types of attachment: rear axle housing, tractor side frame, and tractor front plate. The rear axle housing required clamps, similar to the U-bolt used for attaching the rear wheel fender. The side frame required square shouldered or headed bolts. The front attachment used capscrews.

A comprehensive Implement Attachment Specifications was also provided which included attaching points, performance ratings, system capacities, weight distribution, and tire ratings. Detailed scale drawings of the complete tractor were also included.

Assuring the intended functional relationship between the tractor and the equipment was the assignment of a Product Engineering Center group known as Implement Compatibility. They verified items such as the mounting, range of movement in the field, weight distribution, hydraulic lift pressures, stability, operator vision, and clearances for maintenance or adjustment. Final approval agreement on revisions required reaching mutual resolution between tractor and equipment engineers.

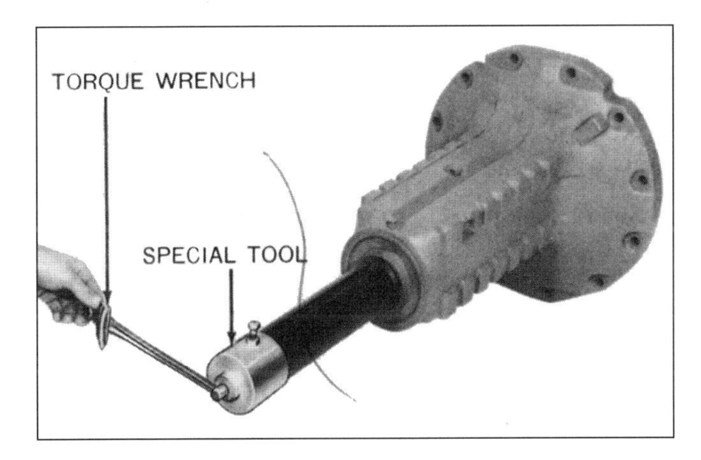

This view of the axle housing shows the notches for the fender and implement attachment.

Final Preparations

Some of the earliest experimental OX and OY tractors began field durability tests and hitch evaluation tests in 1956. Throughout the program the 730 hitch or the comparable towed implement was the usual standard of comparison. Occasionally, comparison included certain competitive tractors with three-point hitches. The performance goals were evaluated with tests such as implement ground entry and exit, uniformity of cut, depth control, hill and swale, and trailing characteristics.

Photographs of final designs were provided for marketing or service groups as production neared. Wilbur Davis, Field Test Project Engineer, gave an account of this procedure in the last hectic days before production:

"In the late fall of 1959 it became necessary for the implement factories to get final equipment compatibility evaluation and approval. Also, photos were required for sales literature so there could be simultaneous release of equipment and tractor advertising. Because of security provisions, this work had to be done either at Waterloo PEC or one of the secure field test sites.

"This program was undertaken with a very minimum availability of experimental tractors and before the last built tractors were available. In general, an updated or latest model experimental tractor was not available. However, there were OX and OY tractors available which were current about the rear end. Therefore, many photos were taken of rear-mounted equipment, including three-point equipment, from the rear only. At times a white sheet or other means was used to hide the front part of the tractor.

"Required field scenes were frequently taken at enough distance so the comparatively minor differences between the experimental tractor and the intended production were indistinguishable.

"There was also much anguish over scheduling of the trials and facilities. The only indoor facilities available at PEC were the old steel building that at times during the winter was very cold. Also, very tight scheduling occurred and many photographers had to work long hours (customarily done only by field engineers).

"Taking tractor and implement field scenes were also painful. Not only did the equipment need to be current and in good mechanical condition but they had to be slick, clean, and green. Moreover, photographers are only 'fair weather' friends. Photos could not be taken early or late in the day because of long shadows. The sun and clouds had to be 'just right,' as well as the background. It was an education for both photographers and engineers to endure these events together."

The U-bolt was required for attaching the rear wheel fender.

Operator Station, Controls, and Chassis

The first John Deere two-cylinder tractors were strictly functional, appearance was not a priority. No particular design effort was made to improve its appearance to help stimulate customer acceptance. A product design decision in the late 1930s suggested that improved appearance might appeal to the customer. If surveyed, farmers, being practical persons , would have said “no” to the idea or “if it doesn’t cost extra.”

The styled appearance was first introduced in 1938 on the Model A and B two-cylinder tractors. Comparing the unstyled and styled models illustrates the functional appearance, yet making them more pleasing to the beholder’s eye. Wayne Worthington, then the Design Chief Engineer, summarized the success in tractor sales of this styling when he

Styling Objectives (as previously stated)

- *The styling should not dictate the tractor design. The styling should be functional.*
- *The tractor should be green with yellow wheels. The occasional use of black was considered permissible on non-prominent parts since the rubber tires introduced the color.*
- *It should be a clean design — not having parts projecting beyond the basic sheet metal.*
- *The rear of the tractor should be simplified in appearance.*
- *Find a better location for the fuel (and LP gas) tank.*
- *The tractor should be more compact for its power.*
- *The underside of the tractor should appear plain, with the frame alongside the engine.*

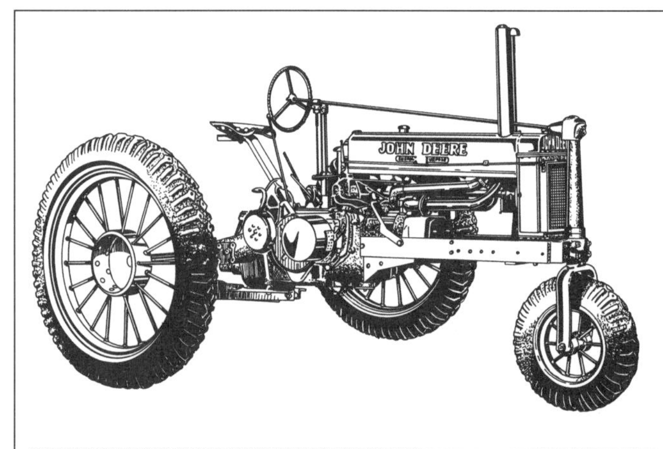

The Model B tractor is shown before and after restyling by Henry Dreyfuss.

Elmer McCormick, the Waterloo Chief Engineer from 1931 to 1948, traveled by train to New York in 1937. He called without an appointment upon Henry Dreyfuss Associates, a reputable industrial designer firm. When Henry Dreyfuss received Mr. McCormick, he felt he needed more background on the business of Deere and Company. He asked his aides to supply him with a Dunn and Bradstreet report on the company. We don't know what either said, but Mr. Dreyfuss agreed to take the return train with Mr. McCormick to Waterloo. And so, "the rest is history."

Since that initial meeting, Henry Dreyfuss Associates has been consulted for styling purposes on most of the products in the agricultural, lawn and garden, and construction lines. Deere and Company believed there was considerable merit in having the development of styling and industrial design performed by outside consultants. The consultants have considerable talent available to them in the field, which can be difficult to achieve with in-house design persons.

said (speaking of the cost at the factory), "that was the best $100 we ever put into a tractor." Following these models, other tractor models in the line received styling. Today, all newly introduced models have included styling for appearance.

Along with improved appearance, the consultants also made advancements in operator comfort and vision and improved the controls. The process typically began with hand sketches, then proceeded to wood, fiberboard, and clay mock-ups. The mock-ups were usually made first in 1/4 scale and later in full scale. They also reviewed the appearance of certain key experimentally made parts and assemblies before they were released to the factory. Of course, there were many revisions and reviews along the way.

Sheet Metal Parts

The most visible parts of the tractor's appearance were those made from sheet steel. These parts included the hood, front plate, engine side panels, grille screens, control support, dash, fenders, and hydraulic function covers. A brief description follows of why each part was designed to achieve function (sometimes more than one function) as well as appearance. A policy rule on exterior sheet metal parts was that no visible spot welds were permitted. All necessary joints in the sheet metal should be a part of the design. They should never appear arbitrarily in the middle of an important form, such as the hood. Latch attachment of the removable parts was concealed, as on the front hood.

Hood

The hoods of the two-cylinder tractors were an assembly of several flat-formed panels. The new hoods were formed in one piece in a large hydraulic press. The rounded surfaces and curved contours gave a pleasing compact appearance. The curved hood form, as viewed from the side, reduced the effect of the hood sloping uphill or downhill when different size tires were used.

The front hood was attached with a concealed latch.

These curved shapes added strength without extensive internal reinforcing. The one large hood covered the entire tractor from the controls section forward. By using the quick release hood fasteners, many tractor assemblies could be quickly seen and were available for maintenance and repair. (When viewing a removed hood lying upside down on the floor, this writer has often been reminded of sections of a canoe. I'm waiting to learn of some enterprising tinkerer assembling a canoe from two hoods.)

A major function of the front plate and front frame support was to provide upper attaching points for front-mounted implements.

Front Plate

The 1/4-inch thick steel front plate was a highly visible part of the tractor. This plate might have been a large casting. It would have some appeal since Deere & Company has one of the world's largest cast-iron foundry capabilities. However, steel is stronger and has a smoother finish. This plate not only is a major part of the tractor's front appearance, but has other functions as well. These functions include attachment for the fuel tank, air cleaner, radiator, oil cooler, and front hood support. It provides a major function in having upper attaching points for front- mounted implements.

Front Side Panels

The front side panels cover the space between the hood and frame, and the front plate and radiator. They direct cooling air flow around the fuel tank. These panels also were quickly detachable for internal cleaning and service. They were formed with a small crown for added stiffness and to avoid a reflective highlight. The front panels were attached to the fuel tank and in turn provided the attachment for the front of the hood.

Grille Screens

The side screens admitted most of the air flow for radiator cooling (and for the air cleaner intake on some options). The screens were made from specially perforated, then corrugated, sheet steel. This concept of perforated and corrugated screen was used on the two-cylinder tractors and was first introduced on the Model R tractor in 1949. The service person could

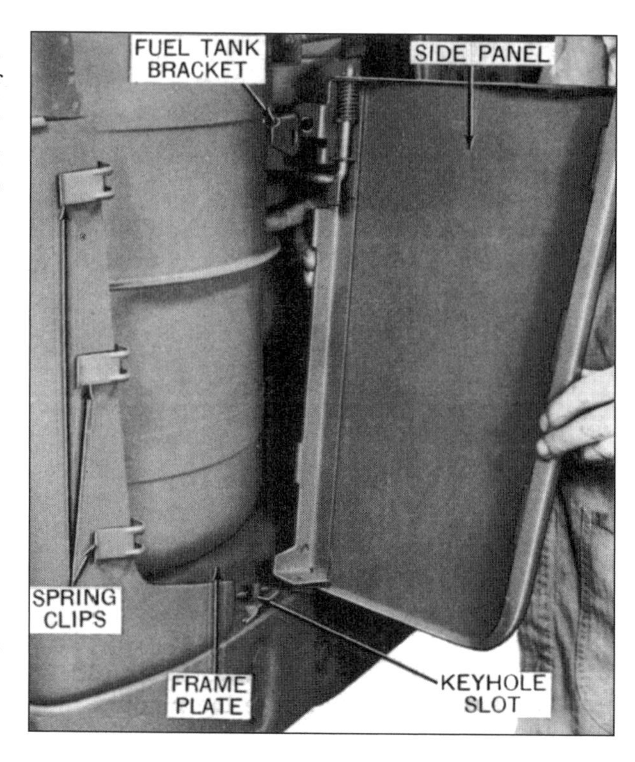

The front side panels were easily detached for cleaning and service.

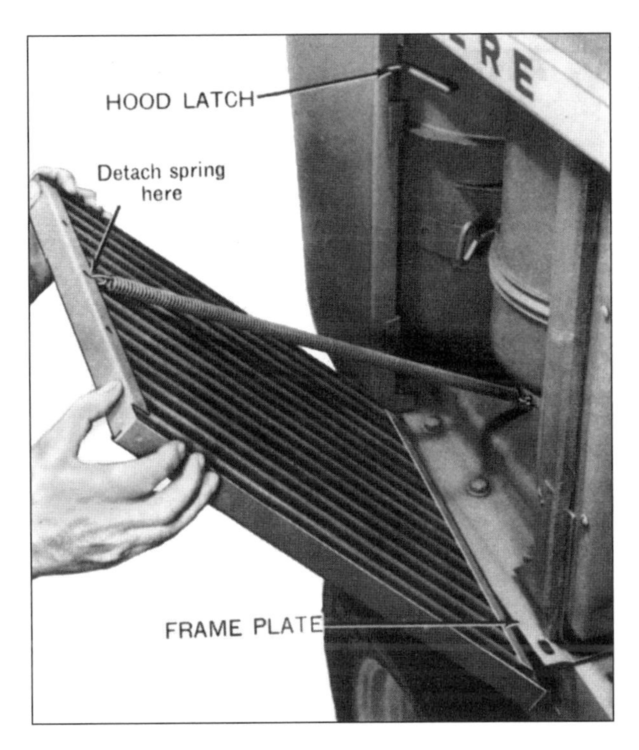

By removing the side grille, access was available to the air cleaner and to the fasteners for the hood and side panels.

easily clean the corrugations by raking his fingers down the screen. While admitting air, the screens excluded large airborne dust and debris particles that could clog radiator cores. These side screens alone provided insufficient air for cooling. Therefore, a small screen was added atop the front plate under the front of the hood.

The screens were attached by tension springs and retaining pins. And they were quick and easy to remove. By removing the side grille screens, access was available to the air cleaner and to the fasteners for the hood, side panels, etc. Thus, the spring-held removable screens were the key to a front appearance showing no fasteners.

Fuel tank

The completely enclosed fuel tank was not visible on the OX (4010) and OY (3010) tractors. (An exception was the OX LP tank as shown on page 88.) It was secured to the front plate and the upper frame plate. In turn, it supported the side panels and aligned the radiator and oil cooler. The fuel tank capacity objective was to provide approximately 6 hours of operating time at the operating maximum power rating:

Gallon Capacity/Hours of Operation

	4010 (OX)	3010 (OY)
Gasoline	34.1/5.95	29.1/7.29
Diesel	34.1/8.75	29.1/10.45
LP gas	39/5.18	25.1/4.77

LP gas, like gasoline and diesel, is a combustible fuel. Since it is stored under pressure and is more volatile, safe care is required. One tractor operator with a field test tractor at Helena, Arkansas, found this out the hard way. While refilling the tank with propane at night, he dropped his glove. He lit a match to find the glove — and the tractor burned.

The 6-hour objective was a parameter from experience at which the fewest customer complaints were received of "small tank" or as more frequently stated "high fuel consumption." This parameter apparently meant the customer was willing under heavily loaded operations to refuel at midday and have enough operating time.

The enclosures were removed to show the fuel tank, air cleaner, and radiator.

Control Support

The control support, a 0.120-inch thick steel part, formed a "U" shape in the horizontal plane. Visible, but not a prominently appearing part, it performed numerous functions. It was mounted on top of the clutch housing and contained many electrical connections. It also supported the rear of the hood, supported the dash, steering, and control levers for the engine, transmission, and hydraulic selective controls. Three screws also attached it to the rear of the engine cylinder head. This support permitted the tractor to be split at the clutch housing, and all the controls could stay with the engine. The control support contained so many hydraulic, electrical, and linkage parts it was humorously called the "garbage can."

Side Shields

The side shields, of 0.060-inch thick steel, concealed the rear hood attachment. During testing of the experimental tractors they were found necessary to direct the hot engine fan blast away from the operator foot area. This situation could create discomfort for the operator if the shields became lost, misplaced, or damaged. Once again, concealed latches were used for attachment.

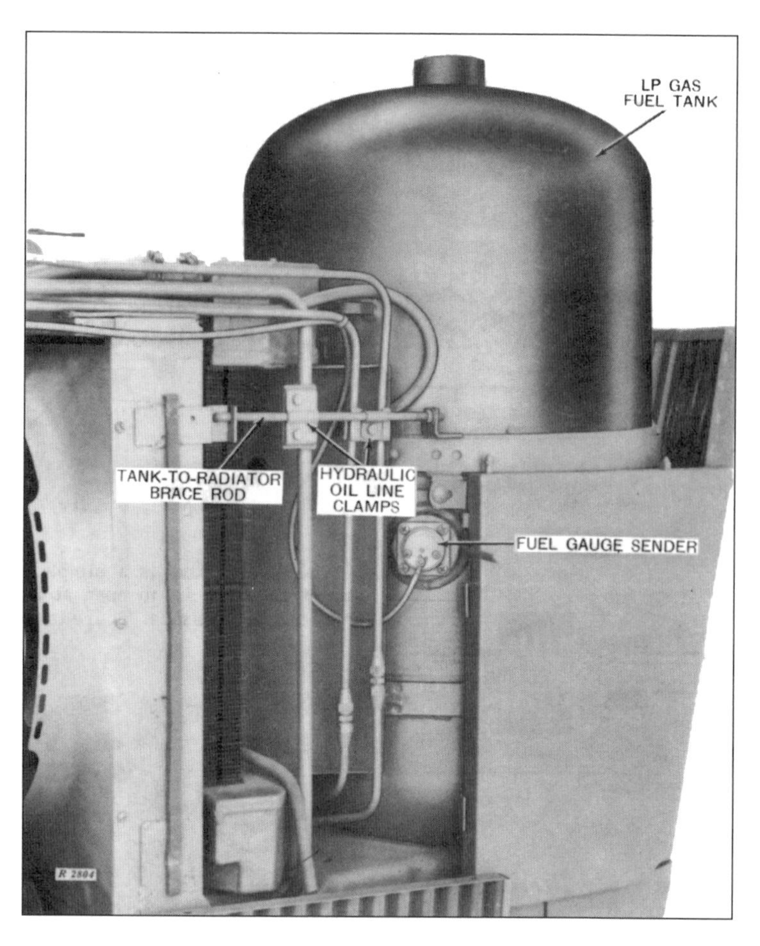

The LP gas tank, unlike the gasoline and diesel tank, was taller on the 4010 tractor.

This right side view has the hood and side shield removed. Observe the control support behind the engine and the "clean" underside of the front structure.

Concealed latches were used for the side shields.

Dash

The 0.060-inch thick steel dash is a highly visible part of the tractor to the operator. It had to be narrow to allow visibility down the side of the tractor, but still contain all the necessary controls and instruments. Obviously, this required careful allotment and arrangement of space for the respective controls and instruments.

Fenders

The row-crop tractor fenders were original with the '30 series tractors. They principally covered a larger portion of the rear wheel than the earlier fenders to provide greater operator safety. The fenders attached to adjustable height brackets to accommodate various tire sizes. They were strong enough that should an operator be thrown against them, they would be supportive. Each fender contained dual head lamps.

The addition of a grommeted hand-hold opening in the front of the right fender was the only change from their use on the two-cylinder tractors. This opening accommodated mounting and dismounting from both sides of the tractors, a feature not possible on the two-cylinder tractors.

The standard tread tractors also used the fenders from the previous two-cylinder standard tread tractors. New small fenders were introduced on 3010 (OYU) utility tractors.

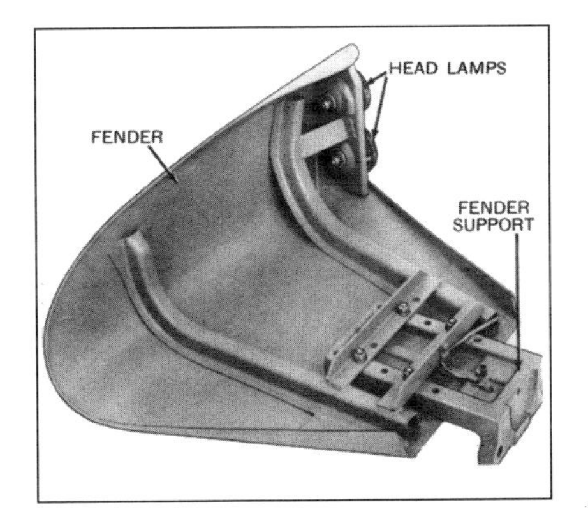

The fenders for the 4010 row-crop tractors covered a larger portion of the rear wheel to provide greater operator safety.

The 4010 tractor dash contained the operating controls and instruments.

Standard tread 4010 tractors used the fenders from the previous two-cylinder standard tread tractors.

Hydraulic Function Covers

The hydraulic function covers may appear to have been only to conceal the castings, rods, and levers underneath. But in addition to improving appearance the 0.060-inch thick steel covers, it also helped avoid operator contact with, or catching on, the sometimes moving concealed parts.

The hydraulic function covers on either side of seat and rockshaft housing prevented the operator from contacting the sometimes moving parts.

The Seat

The seat, steering wheel, and platform are in continuous contact with the operator. Of the three areas, the seat assembly has more affect on the operator's comfort and security of control. The placement of the seat on the tractor and the relationship to controls was made in consultation with Henry Dreyfuss and Associates. The Deere & Company Product Development Department, headed by the late Charles Morrison, designed the initial seat cushions and the suspension system.

William F.H. Purcell described the seat (Purcell, W.F.H., *Automotive Industries*):

"The seat adjusts to the physical proportions of the driver. Resting on an inclined track, the seat slides up and back for taller drivers, forward and down for shorter drivers.

"A quick-release lever jumps the seat backward when the driver stands at the wheel or dismounts. When he sits down again, a built-in 'memory' device returns the seat to the same pre-selected position on the track.

"The seat assembly consists of three separate cushions:

- *a seat cushion,*
- *a cushion for the lower back (or lumbar region) that fuses with cushioned arm-rests,*
- *an upper back cushion.*

"This unique three-part design supports the natural S-curve of the spine. Particularly vital, the lower back cushion minimizes familiar 'low back pain.' The upper back cushion unit is spring-mounted. This gives support to the back on smooth terrain when the driver can lean back. On rough ground he sits up straighter and can avoid any pounding effect caused by the high seat back.

"The cushioning itself is 'variable density foam,' firm in some areas, soft in others. This is particularly important in the seat cushion, which is firm in the center (where the body's weight rests on the bony points of the pelvis). But it is soft at the forward edge, to lessen pressure on the large blood vessels in the thighs.

"The seat suspension is 'rubber in torsion' and has about 4 inches of vertical 'travel' to soak up road shock. A knob permits the driver to adjust the suspension to suit his weight."

The suspension system was complicated by the need to fit around the rockshaft housing and still provide the desired functional features. The seat retraction feature was later compromised when cabs were provided.

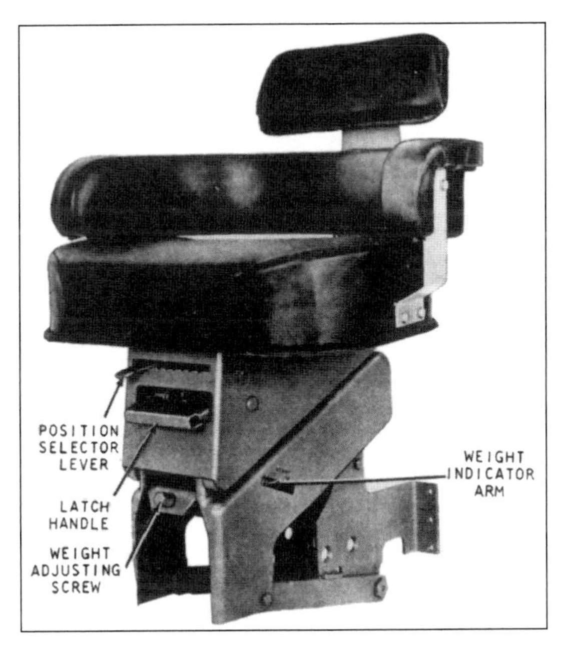

The deluxe seat provides three separate cushions: a seat cushion, a cushion for the lower back that fuses with the cushioned arm rests, and an upper back cushion.

This was permissible since the capability of operating while standing is lost with cabs.

The cushions were designed with additional contribution from Dr. Janet Travell. (Later, she achieved publicity as a consulting physician to President John F. Kennedy on his back problem.)

All the ergonomic changes provided a new seat form, the relationship of the seat, platform, and steering wheel, the steering wheel angle, and some of the control positioning. This design's long production attests to the satisfactory ride it gave the operators.

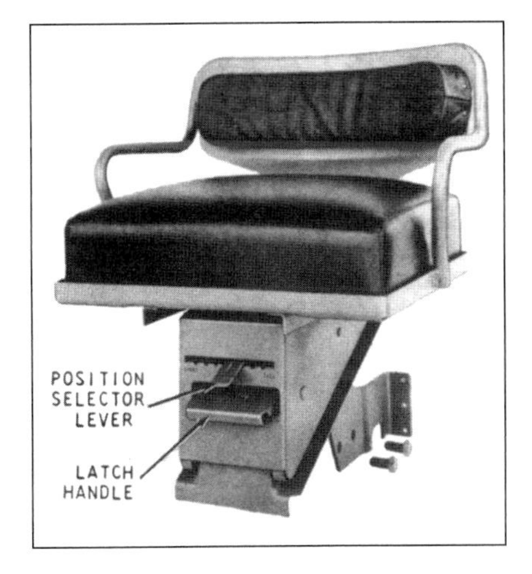

The seat assembly has the most affect on the operator's comfort. The regular seat is pictured here.

Service Instruction Cards

A concern during the experimental program was to provide more "user-friendly" service and maintenance instructions. The reasoning was that better maintenance might result, which in turn would increase customer satisfaction.

A service instruction "mini manual" on plastic cards was accessible through a door in the dash. Occasional updating was possible by an easily removable card feature. Experience with customers and their production tractors revealed the cards were simply not used. Consequently, these cards were eliminated during a later cost-cutting campaign.

Electrical System

Optional electric lighting and starting had been increasingly provided on tractors over the preceding 20 years. By the time of the OX and OY design, it was a given objective that these systems would be standard equipment.

A 12-volt electrical system, with a 12-volt generator, was used for all 3010 and 4010 lighting systems. The spark ignition engine used a 12-volt starting motor and a single 12-volt battery. Because of the need for higher cranking torque, a 24-volt starting motor was used on the diesel engine equipped tractors. To achieve this need, two series connected 12-volt batteries provided the 24-volt starting current. Charging the two batteries was done using the 12-volt generator connected in parallel to the batteries. It was problem-free if the battery was used for only starting. However, 12-volt lighting was also needed and used.

Several alternatives were considered to provide lighting with the 24-volt starting. One option was to use 24-volt bulbs for the lighting systems but they had more fragile elements than 12-volt bulbs and therefore had a shorter life. Another idea was to use 24-12 volt crossover switches which were commonly used for starting heavy-duty construction equipment. These crossover switches were not considered for these tractors because they were cumbersome and costly.

The ultimate solution was to divide the lighting circuit into two nearly equal current requirements. Each circuit side took current from one of the batteries. The lighting current requirements could not be made completely equal, nor was the usage of the lights always equal. Hence one battery predictably over time would become more discharged than the other, would not get recharged as much, and gradually became able to hold less charge. This unbalance of the discharge-charge cycles had been in use on the 730 diesel tractor, and was considered unsatisfactory but tolerable. This arrangement continued in production with Deere diesel tractors until higher torque 12-volt starting motors became available. These newer 12-volt motors were comparable to the 24-volt motor in physical size.

The 4010 diesel tractor electrical system required a 24-volt starting motor. To achieve this task, two series-connected 12-volt batteries provided the 24-volt starting current.

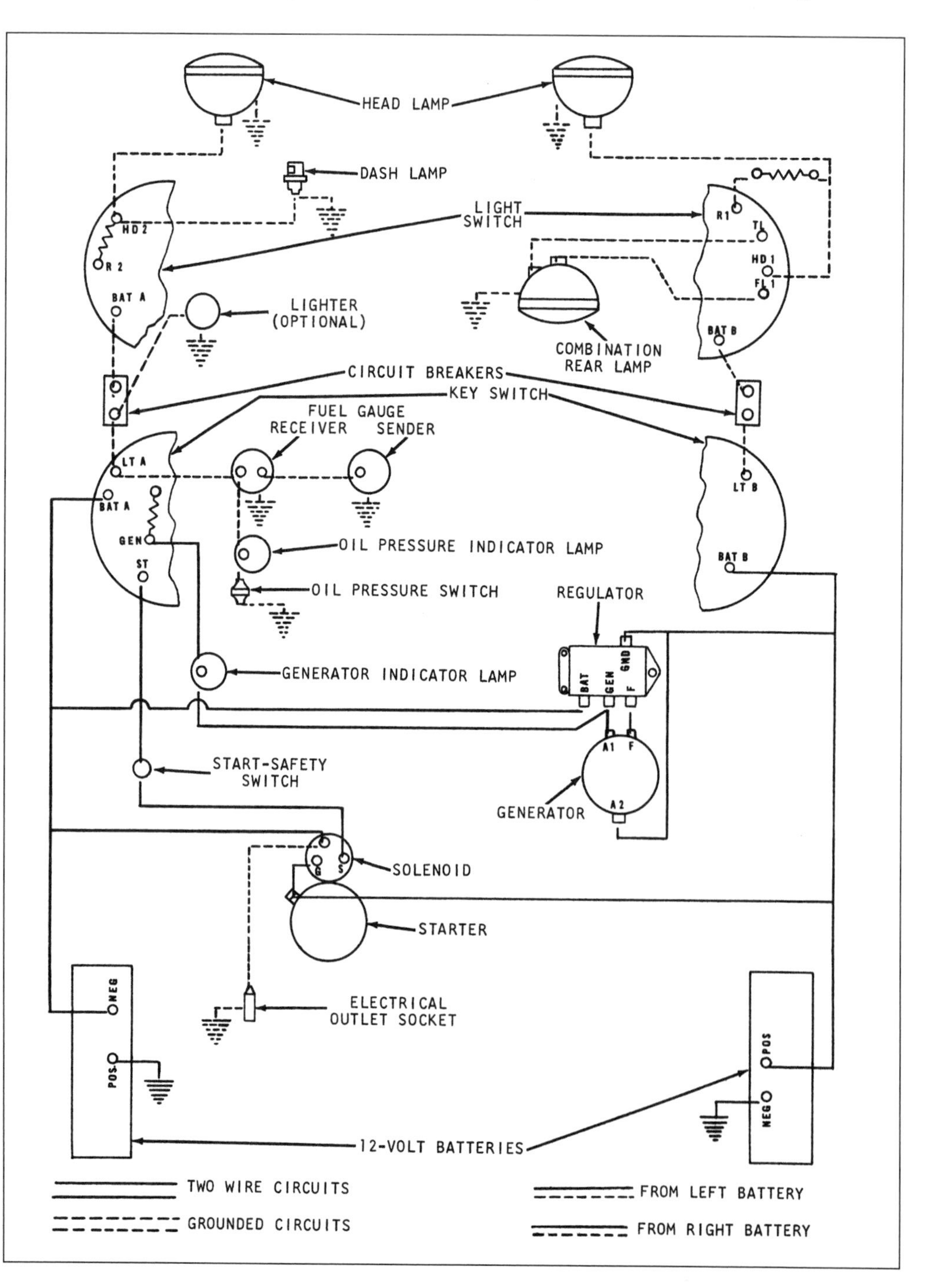

Frame

Broadly described, the frame of a tractor is the total structure from the rear wheel axle attachment, through the various castings, to the front wheel attachment. This structure may appear to be a rigid unyielding body that wards off all external impacts. However, each axle shaft, casting, and steel side frame has a given bending deflection and stress characteristic. (Tests in the 1970s found that the total structure is flexible enough that on highway travel the rear tires can find a common dynamic frequency. Then an uncomfortable ride results for the operator.) Stiffening of the structure was helped by using the engine block to supplement the small torsional stiffness of the structural steel side frames (see page 88). The engine stiffness was increased because of these imposed loads and the potential for damage to internal engine parts. For instance, a cast-iron oil pan was used instead of the first-tested stamped steel oil pan. These changes made the engine heavier and more costly, though still it was a cost justified from the overall tractor standpoint.

The side frames were 5 inches wide, 9-pound-per-foot structural steel channels. They had punched holes as part of the implement attachment configuration. Their stiffness was also improved by upper and lower 0.120-inch thick steel frame plates. The lower "X" plate was used only on the tricycle tractors. The upper frame plate also aided in attachment of other front-end parts and had an embossed seat for the fuel tank.

During the OX and OY programs the frame structure stiffness was tested in the field. It was also tested in the laboratory, under torsional loads approximating that imposed by front cultivators. Simultaneously, various wheel loads were also imposed. These laboratory test fixtures used hydraulic cylinders to apply the respective forces. The fixtures usually ran 24 hours, 5 to 7 days a week. Here again, recent modern stress and computer analysis has shortened the time to develop a satisfactory frame structure in ways not available in the 1950s.

Radiator and Cooling

At Deere Waterloo Product Engineering, the chassis design and test groups were responsible for engine and vehicle cooling. Obviously, coordination with engine, transmission, and hydraulic requirements was a given.

The choice of a front-mounted fuel tank has already been discussed. Some may wonder — did the fuel tank hamper adequate radiator cooling? The answer is yes and no. "Yes," the radiator was not open at the immediate front to an unobstructed grille for incoming air. "No," other ways of introducing sufficient outside air were provided.

The side grille screens were the most noticeable way to provide airflow. The air passage size was limited horizontally by the radiator proximity to the wide part of the fuel tank. Vertical limitations included the frame and hood lines.

Deere's positive experience with the convoluted perforated screen suggested its continuation with the new designs. The depth and width of the convolutions, as well as the number, spacing, and size of the perforated holes, have been developed with a purpose. These convolutions theoretically provided a total hole area theoretically equal to the grille frame opening. Air velocity through the holes was low enough that orifice restriction allowances were not a factor in this empirical design.

The proof of the cooling system adequacy was primarily a test of the loaded tractor in a temperature-controlled room that had controlled air circulation. As the room, or ambient, temperature rose, the temperature of the radiator coolant also rose. Design, test, and customer experience dictated the ambient temperature for which adequate cooling must be provided. Laboratory testing goals included some built-in allowance for actual field conditions of less than clean coolant systems, some plugging of grille screens, and downwind tractor operation.

Steam may occur at the upper limit on coolant temperature. This temperature limit may be slightly higher with pressurized radiator caps. These subject the whole system to higher pressures, making radiator strength a factor. Careful attention was given to preventing steam formation in isolated or poorly circulated pockets in the engine part of the coolant circuit.

Early cooling testing of the new tractor designs showed that not enough air was permitted through the side grille screens. A smaller screen was added above the front grille plate to correct the problem. It provided air flow over the top and around the sides of the fuel tank toward the radiator. Most large Deere tractors needed these screens.

As explained in previous chapters, a transmission and hydraulic cooler were added to the new tractor design. This small oil-to-air heat exchanger was placed in front of the radiator and behind the fuel tank. The oil cooler covered part of the radiator front opening.

The engine air cleaner was also in the compartment with the oil cooler. Air was expected to flow over its round shape with minimal restriction. The foam sealing strips, although unnoticeable, were an important part of the cooling system. These strips sealed the perimeter of the grille screens, the upper frame plate, and the hood to the radiator top tank.

A successful cooling system is subject to many factors such as the effects of incoming ambient air and the oil cooler which have already been discussed. Other factors are the engine power, rate of engine heat dissipation to the coolant, water pump capacity, and connecting hose sizes. Radiator factors include top and bottom tank sizes, radiator core fin spacing and design, and sensitivity to debris.

Front End Weights

Restyling the front end weights became an objective of the new design. Part of the goal was to reduce the unit weight for easier installation and removal. Wendell Van Syoc related the display of the proposed design at a demonstration for management:

> *"Cast weights were not available for the scheduled showing. Weights were made of wood and painted to show the new improved design appearance and method of attachment. As the master of ceremonies described the merits of the new weights, Ken Murphy removed weights, one by one. He carefully simulated handling 85 pounds of cast iron as he lowered each weight to the ground.*
>
> *"After the demonstration, the clean-up crew moved in. Leonard Jorgensen tucked one weight under his arm, and with a weight in each hand carried three weights out of the demonstration area. The guests had a good laugh as Waterloo personnel informally told them the rest of the story."*

Operator Protection

Soft canvas enclosures had been available for two-cylinder tractors for several years. The enclosure helped warm the operator during cool or cold weather by channeling some engine heat to him. Merlin Hansen encouraged the evaluation crew to investigate other methods of warming the operator. Means for warming the platform and the seat area were made using engine coolant for the heat source. They were not very successful, as these did not warm the operator sufficiently enough in severely cold weather.

The 3010 and 4010 tractors were introduced when large amounts of corn were still being harvested by pickers. Wilbur Davis related:

"One of the requirements for the new tractors was adapting to the mounted corn picker. Guidelines were that the attachment should be similar to that used on the two-cylinder tractors. Adequate operating clearances, and additional provision for the tractor cooling air was needed through the side grille screens. Corn pickers generate much corn leaf and tassel debris that becomes attracted to the tractor cooling air inlet. A 'rabbit ear' design was provided which gave vertical inlets that mounted in place of the regular grille screens.

"Test of the special vertical air inlets was considered most critical early in the season when the ambient temperatures were highest. One test day, Ken Murphy and I were testing an OY tractor when a fire was found on the side of the tractor. If it had spread, the experimental tractor would have been lost, plus the cornfield. Fortunately, by using 'quick and expedient' means, the fire was extinguished."

The Deere & Company Product Development provided an experimental cab that met the clearance requirements of the mounted corn picker. The whole front of the cab was hinged from the top and swung forward and up to allow operator entry and exit. This cab was ahead of its time — nearly 10 years before cabs were available on Deere tractors for customers. Again, quoting Wilbur Davis:

> *"It was an impressive experience, and an insight into the future to operate an experimental tractor and a number 227 corn picker on a miserably cold day in Iowa. The operator could be in shirt sleeves inside the cab."*

The Roll-O-Matic was an optional front axle attachment.

Version Models

Various crop and regional requirements necessitate different wheel tread spacing, tire and wheel equipment, axles, and fenders. Attachments to the basic tractor provide cost-effective measures for the desired configurations. The basic design configuration was the general purpose, sometimes called tricycle, tractor. Optional front axle attachments included the Roll-O-Matic, the double knuckle (fixed axle for two close front wheels), and a single front wheel. With the adjustable tread (wide front) axle option it was no longer termed a tricycle tractor. A selection of front tire, rear tire, and wheel weights was also available. And as described earlier the flat-topped added-coverage rear fender was used.

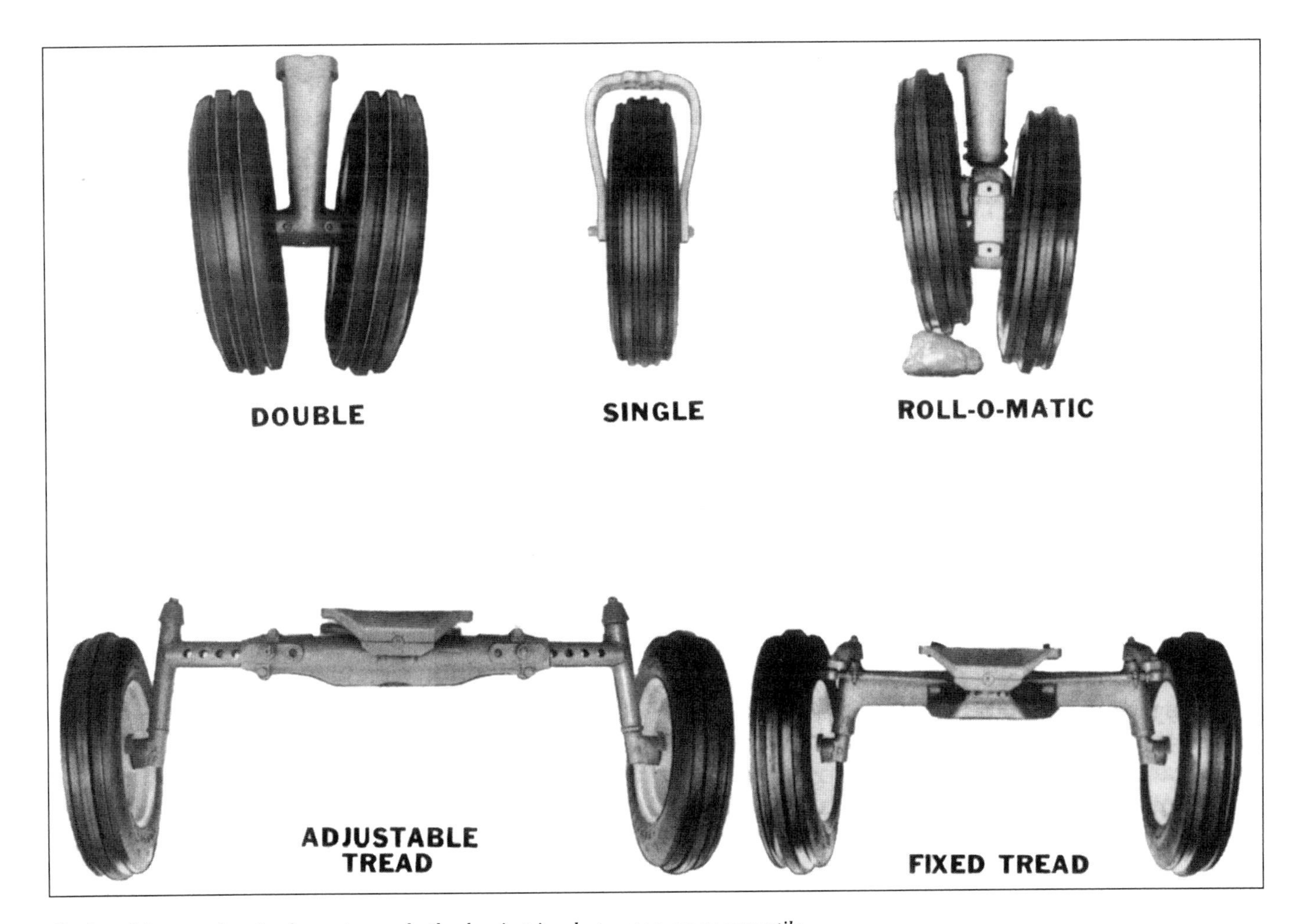

Optional front axle attachments made the basic tricycle tractor more versatile.

This 4010 hi-crop tractor with a gasoline engine was another version of the 4010.

Changes to provide other versions of both the OX and OY included:

- Standard tread versions. These tractors used adjustable tread and fixed tread front axles and different fenders (modified 630 and 730 standard tractor fenders). Also, the battery boxes were moved to between the platform and fender, and a selection of various tire and wheel combinations was available. The shorter wheelbase on the standard tread tractors was achieved by reversing the front axle assembly in its pivot bracket.
- High-crop versions. These tractors continued assemblies from the 630 and 730 tractors. They included the drop housing final drives, large size rear tires, and high-mount front axles or a single front wheel.
- Utility version.
- Orchard version.

3010 (OY) Utility Tractor

More modifications were necessary to create the special design OY utility tractor. The objectives for this tractor included a short length, a small turn radius, and low down operator position.

The short length was accommodated with confining the availability to only the OY. This was, of course, two engine cylinders shorter, and had a shorter and narrower transmission case and clutch housing. Two wide front axles were provided. One (called the "swept back"axle) moved the front wheels and steering attachment rearward for even smaller turning radius. This was done by reversing the front axle within the frame attachment. This procedure reduced the wheelbase from 92 3/4 inches to 81 1/2 inches.

The operator was moved about 4 3/4 inches forward and 8 inches lower than in the row-crop version. As a result, another new special seat suspension with less travel and lower adjusting ramp angle (from 25° to 19°) was required. The 45° angle of the steering wheel was changed to

The objectives of the 3010 utility tractor were a short length, a small turn radius, and a low down operator position. The 3020 utility tractor with a gasoline engine, pictured here, was very similar to the 3010 utility.

65° angle from the horizontal, and placed closer to the dash. This change prompted the hand speed control lever to be re-oriented so its travel was parallel to the top of the steering wheel rather than at the side. The gear shift lever was shorter for clearance with the closer steering wheel proximity. The operator sat astride the transmission case because of the lowered seat position. New right and left separate platforms were used instead of one wide platform.

Orchard Tractor

The orchard tractor was an important but low volume tractor. It was constructed by adding the appropriate shielding and fenders to the 3010 utility tractor.

Function and Appearance

Various major chassis parts were designed for multiple functions and for cost effectiveness. They were also scrutinized by the industrial designer for appearance. As demonstrated on the two-cylinder tractors, meeting styling objectives improved the customer acceptance of the tractor. Subsequently, it was not considered necessary to change the basic styling for 12 years (1960-1972), which testifies to the acceptability of the concept.

This 3010 diesel orchard model was pulling a disk harrow.

Development and Testing

Engineering tests of agricultural tractors (both experimental and production) fall into four broad groups using laboratory and field facilities for both durability and performance. Wilbur Davis, Field Test Project Engineer, emphasized the importance of this aspect of a new tractor program when he said:

> *"I believe there is a significant story that the tractors' success was due to splendid specifications and design, and adherence to a comprehensive proving program. The latter includes durability testing criteria, disciplined field evaluation, good development analysis, a strong field testing program, and a structured implement compatibility program."*

"Test work often also includes trial of improvements that may be shown as needed. Deere calls the combined task of redesign and retest — development."

1. Laboratory Tests that Help Determine Durability

Examples abound of the life testing of engines, transmissions, hydraulic assemblies, and chassis structural assemblies. Satisfactory life objectives under certain laboratory applied loads are established by the particular engineers involved. Special test fixtures may need designing and fabricating. These tests may operate 24 hours a day with shutdown only for maintenance. One laboratory employee usually oversees several similar durability tests.

2. Laboratory Tests that Help Determine Functional Performance

Examples of these tests would be the power output of an engine, the breaking strength of a shaft or lever, and the noise generated by the tractor or an assembly. Other examples include the natural frequency and amplitude of the vibrating tractor or assemblies.

3. Field Testing for Durability

The procedure most commonly used by Deere & Company was to place the test tractors with large operators who could use them inten-

A laboratory employee usually oversaw the durability tests, like this 4010 diesel engine dynamometer test.

The tractor or components being tested were placed in the cold room for overnight cool down before the tests. Instrumentation was attached. The usual cold extremes of –20° to –30° F were controlled.

sively. The goal was to gain as much as 2,000 hours per year per tractor. This compares with the average farmer's use of 300-600 hours per year. On production tractors this aggressive testing may uncover a deficiency and establish a correction for production and for customers before major customer tractor failures occurred. Experimental tractor programs have deadline commitments. It was important that the durability be known before committing to the next important activity toward the start of production.

Some durability testing was performed on the agricultural land owned by Deere adjacent to the Product Engineering Center (PEC). Most field durability testing could be performed in large land operations away from Waterloo. These farm operators provided the fuel, drivers, field work, shop and office space for the Deere test representative(s). The Deere test supervisor and his staff maintained the tractors, changed test parts, kept parts records, fuel and oil consumption, and hours of use. The tractors were used by the farm operator in his fields as he would his own tractors.

To test for durability, Deere & Company typically placed the test tractors with large farm operators who would use them intensively in the field.

Most such operations were in the south, southwest, or Pacific coastal states because of the longer periods of favorable operating weather. However, if information on durability characteristics in special regional areas was required, additional test sites were established. Examples may be wheatland farming in Montana, Colorado, Oregon, Washington, or California; rice field farming in Arkansas.

One of the benefits of field durability testing on actual operating farms was the type of driver exposure. These drivers were usually unconcerned about the test results and untrained in the technical work under way. Yet, tractors need to be designed to be understandable by unskilled drivers and be durable if modest mishandling occurred. At Laredo, a typical driver might be an illegal Mexican immigrant. When a failure occurred it could be difficult linguistically to get a good description of the happening. Wilbur Davis related several instances:

Tractor rolling load testing permitted controlled durability testing of the tractor engine and transmission when environmental conditions prevented actual field testing.

"Something was wrong with the tractor, and it stopped. The driver was apparently so afraid of some repercussion that he waded back across the Rio Grande River and hasn't been seen since.

"In Arkansas, a tractor ran into a tree near the test site shop. The driver was asked what he was doing when it happened. He stated that he was bending over to operate the brake pedals by hand.

"Also at the Arkansas test site one very good driver was deaf. One afternoon he brought in a tractor saying it had an unusual vibration. The shop crew investigated and found that a wheel was loose from the rim.

"Again, at the Arkansas test site, one tractor was mostly lost by fire. Reportedly, the operator was fueling the LP gas tractor and allowed the gas in the top of the tank to escape when a lighted cigarette was present. The heat from the ensuing flame caused the relief valve on the tractor tank to open and then the flame became 50 or so feet high."

4. Field Testing for Evaluation of Functional Performance

Tractor observations in actual field operation is important. Testing should be with a variety of implements and in a mixture of soil, terrain, and crop conditions. For example, steering a tractor with an implement on an Oregon sidehill presents different requirements than operating it in muddy Arkansas rice fields. Trailing characteristics of a hitch-mounted implement could be different on an Iowa sidehill compared to over-hill-and-through-swales also common in Iowa. Often field tests were made where the tractor pulled only rolling dynamometers. Maximum use was made of the Product Engineering Center fields and crops, and nearby (this may mean as much as 40 miles away) private farm sites.

Farming in Iowa is highly seasonal because of the weather. Test programs were also usually organized for winter operations in the south or southwest. The opportunity to conduct the tests away from the interferences of the engineering office allowed an often ambitious schedule to be expedited and achieved. Product Engineering Center engineers, and other Deere factories' engineering test persons, were present for com-

This photograph shows the field testing of a tractor towing a dynamometer truck. The rear unit was a modified "tractor" which could provide variable loading of the test tractor. The modifications changed the engine to an air compressor. The transmission could be shifted to vary the load level.

Many engineers visiting the test sites found that their suitcases became the travel case for new or redesigned parts going to the sites, and damaged or worn parts for the return to the Product Engineering Center.

parative tests of their new equipment. The evaluation crew members gained overall exposure to the field operation. The screening of various design options, and acquaintance with many persons outside the Product Engineering Center, were additional benefits. This list could include component suppliers, implement factory personnel, Deere & Company personnel, drivers, and test site personnel.

Tests with competitive tractors and implements were an important comparative technique. No standardized format for the tests existed. The format of previous testing was revised as necessary to observe a new feature or circumstance.

One of the measurements taken by the evaluation crew was the depth of the plowed furrow. It wasn't just a one-time check. About 25 readings were taken of each pass of the field of each model. This information was then used to calculate a statistical average value. Wilbur Davis related:

> *"This was a tedious and time consuming task requiring two persons. One measured the depth and the other tabulated the reading. The scientific measurement tool was a flat board with a measuring rod readable at eye level. Once, two of the Plow Works mechanics, sometimes called the 'gold dust twins,' begrudgingly accepted the task. However, they counted the total and reached 3,423 readings for the day."*

Being a part of these winter test programs allowed the participants to become better acquainted; especially with people from other factories. The days were usually long, and could be hot and dusty, so afterward a shower at the motel was welcome. Later, dinner at one of the local restaurants provided further opportunity to recount the incidents of the day (or other days). Many stories of the days events were told. Intermingling people in this social atmosphere promoted camaraderie and understanding. While this social side of the activities may appear unnecessary, it did contribute to greater cooperation, and as a result, better tractors, implements, and their combination.

Demonstrations

A special part of the winter test programs included a presentation to interested management personnel. Not only Waterloo engineering managers, but often managers from the Waterloo factory and other Deere factories, and some Deere & Company staff were present. Part of the presentation included a comparative driving test of Deere production, experimental, and competitive tractors. A typical agenda provided three sizes of tractors, mounted integral equipment, and towed equipment. The demonstration site would have driving stations and the participants rotated according to a schedule. Each person drove each unit 12 to 20 minutes for some familiarity with each set of equipment.

The equipment was prepared prior to these demonstrations. The preparation included comparable ballasting, implement sizing, and wheel treads. Familiarity speeded this preparation on Deere tractors, but more time was needed on the competitive tractors. Since each design often took a different form to meet a given goal, this compounded the task of getting them ready for a fair comparison. Operating the mix of competitive tractors also caused the visiting drivers some confusion. Wilbur Davis commented:

"The late Lyle Cherry, Marketing Manager of the Waterloo Tractor Works, with a unique flair, was a strong and vocal supporter of the total program. In one instance he brought a competitive tractor and its plow back to the headland. There, with the several participants and test crew present, he crossed himself before getting off that particular competitive unit."

Each driver was provided booklets which listed statistics on the various equipment units of the horsepower, ballasting, and implements. The data often was not complete until a few days before the demonstration. This timing meant that the last person to leave PEC (publishing facility) carried enough quantities of the last revision to the test site.

Some invited participants were uncertain of their driving skills yet wanted to experience the tractor's performance and operation. Then, an evaluation crew member would ride along to coach them.

A summary presentation of the test program findings was usually provided the senior management. Here again, the opportunity to meet and work with various senior management was helpful in promoting good interpersonal relations. It also helped to keep the program on the always varying target of customer needs.

The Texas security test site was approximately 20 miles from Laredo. For one demonstration, it was planned to get a prepared box meal from Laredo by 11:00 a.m. The meal arrived at the test site ranch house on time, the participants arrived, the meal was served — but there were no eating utensils. It was obviously too late to arrange for others to come from town. Improvisations were made by all.

Kenny Murphy, Field Test Engineer, told of a similar situation when Ray Brandt was trying to get the plastic tableware, and called to say he was delayed by the Border Patrol. His car was in the process of being repaired by George Brown, Instrumentation Engineer, and somehow the license plate became covered. Fortunately, the Border Patrol finally accepted Brandt's explanation.

Implement Compatibility

Unless an implement is self propelled, i.e. has its own propelling power train and operator's station, it usually performs its function with a tractor. This enables the implement to be purchased separate from the tractor. It also permits the more expensive tractor to be justified by being used with a variety of implements and operations. Lastly, an implement may receive less use per year than a tractor, and be designed for a different life cycle, with a savings for the customer.

The compatibility of attaching a wagon to a tractor drawbar is much less complex than attaching a front-end loader. However, even the drawbar is standardized in pin hole size, and its location from the tractor wheels, and PTO drive. For more complex mounted equipment, special attaching points are placed on the tractor. These points are not always interchangeable between tractors, and not with competitive tractors. The implement manufacturer wishing to have its product fit several makes and models of tractors needs to supply a bundle of mounting parts for each respective tractor. And of course, a trial fit and operation of the respective implement and tractor are essential.

The 3010 and 4010's implement attaching points are grouped in these categories:

- Drawbar configuration
- PTO shafts configuration and power
- Hitch configuration and lift capacity
- Tire and weight availability
- Rear axle attaching points
- Side frame attaching points
- Front plate attaching points
- Hydraulic system capacity and pressure
- Tractive power and weights

The drawings and tabulated dimensional data for the above characteristics were grouped and provided on an "Implement Attachment

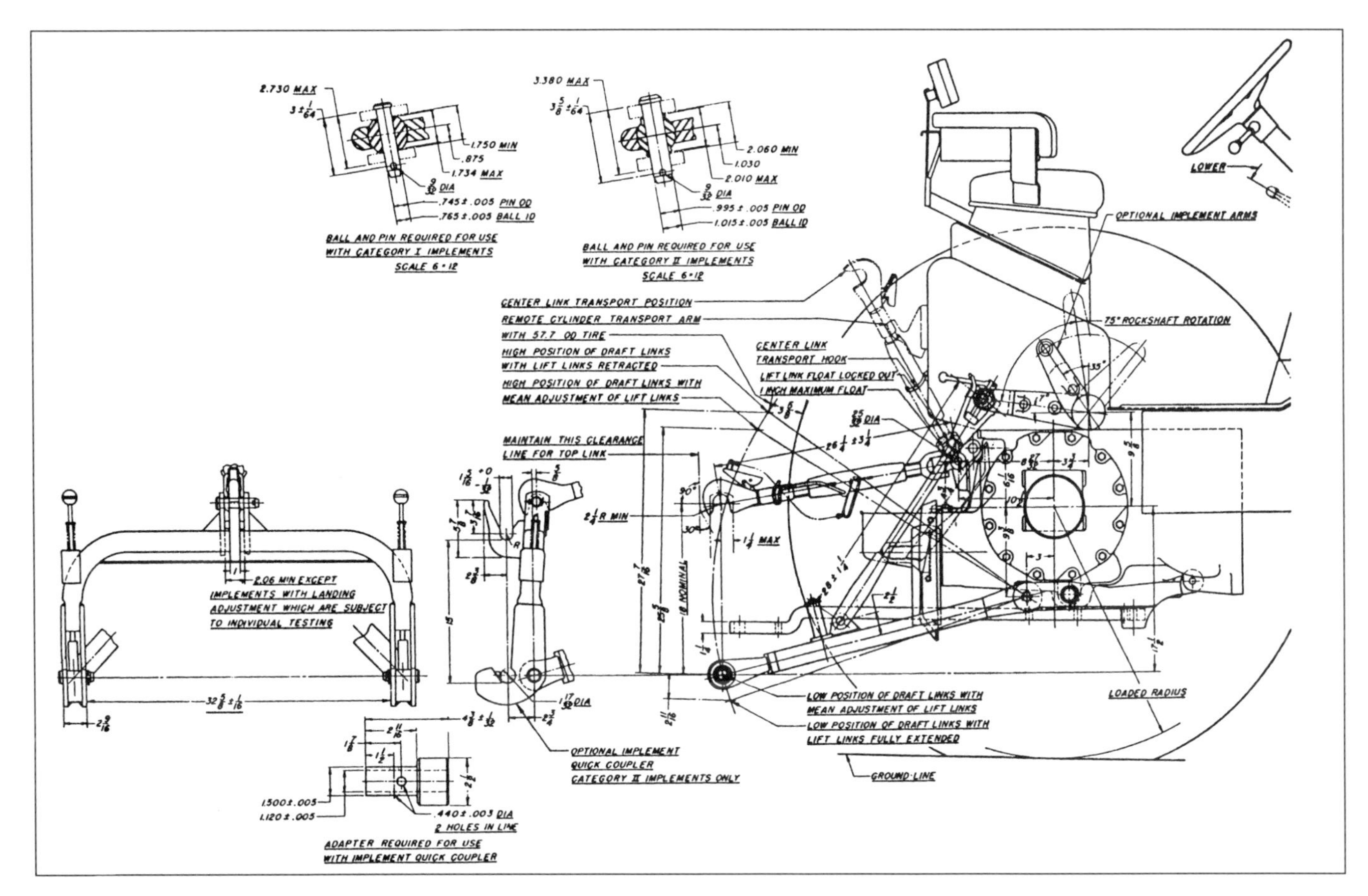

This drawing is an example of the "Implement Attachment Specification" for the hitch and quick coupler.

Specification." Most information on attaching points was placed on a series of drawings of the various tractor views and models. As test work progressed and as the design was revised, updates of the information were necessary. These specifications were made available during the experimental program to other Deere factories for their respective design use. After the tractors went into production, other manufacturers could obtain copies through the factory Marketing Division.

Shipping Tractors

Experimental OX (4010) and OY (3010) tractors were shipped to and from field test sites in closed vans. Initially, the tractor was disassembled to the basic frame and crated. At the site the test crew had to reinstall the rear axles, the front pedestal, and the wheels.

Carefully coordinating the completion of building experimental tractors was planned to coincide with the field test schedules. Adhering to build schedules was required because other major events were scheduled during field test planning. These events included cooperative work with implement engineers, performance evaluation of new design components, or demonstrations for Deere management and other personnel.

Wilbur Davis told of the conditions of receiving a tractor at a test site:

"One time, a major event was scheduled in Texas for the middle of a given week. The assembly and short test of two of the experimental tractors were only completed Friday of the previous week at Waterloo PEC.

"The tractors were loaded promptly and two drivers were provided for the truck-tractor. By Sunday morning the load was in

Laredo. The evaluation crew escorted the truck to the test site and through the locked gates. Other experimental units had earlier been removed from the unloading area. After the truck operators were taken to the gate, the evaluation crew opened the trailer. They found a pile of snow with the tractors. This occurred with shirt sleeve weather in Laredo.

"The test and evaluation crew promptly unloaded the trailer. By then there was too much experimental equipment around to allow the drivers back to move their own truck. So, the test crew drove the truck out to the gate.

"By Sunday evening, the crew had all the tractors assembled, ballasted, wheel treads adjusted, and checked out." (And the final preparations scramble began to prepare for the midweek event.)

Test Security

At the OX and OY Texas test sites, special security procedures were established allowing only qualified persons to enter.

Wilbur Davis recounted: "The test site supervisor for many years was Earl Speer. The field test site staff was instructed to admit only persons with formal authorized clearance. Clearance was usually authorized by certain staff members of the Product Engineering Center."

Most of the test site work was conducted behind locked gates. Occasionally there was a curious person who either climbed the fence or stopped outside to watch. Only in one instance was a person asked politely to leave the Deere property. Usually, the evaluation crew acknowledged the presence of a person, but continued their work without inviting any conversation with the onlooker. The unwanted guest either tired of this treatment, or got the message without gaining much new information.

The side profile of the OX and OY was similar to that of the then-current Massey Ferguson row-crop tractors. The most obvious similarities were the forward projection of the tractor front ahead of the front axle and the forward placement of the operator's seat.

Paint color helped further security of the experimental tractors. Previously, experimental Deere two-cylinder tractors had been painted a khaki green. The public had become aware in test areas that this color was used on Deere experimental tractors. Completing the false identity, the OX and OY tractors were painted in Massey Ferguson colors of red with yellow wheels. They were usually in a field with several competitive tractors, as well as the then-current production Deere models. It is believed that the tractors were field tested from 1956 through 1960 without competition knowing the specifications or performance of the new models.

Wilbur Davis related details of life on one test site:

"Testing was conducted largely behind locked gates on special farm projects. The principal one of these was at Laredo, Texas, on approximately 1,200 acres that fronted on the Rio Grande River and had two locked gates. There was always much discussion on who was to ride in the front passenger seat of the work car because that person was expected to perform the burden of opening and closing the locked gate.

Most of us have seen how a truck driver shears off the top of his semi-trailer when he misjudges the clearance of the bridge he is passing under. A similar experience nearly happened to the writer while observing the test operation in Montana.

The test crew wanted to remove the experimental 5010 size tractor, towing a wide field cultivator, from several miles north of Great Falls to another site several miles south of Great Falls. The machines were moved on little-traveled country roads and it was also necessary to go through the city of Great Falls on some of the side streets. The writer was offered the opportunity to start the driving, while the test crew followed in a pickup truck.

All went well for several miles. I was relishing the experience of driving this large rig down the road at about 20 to 25 miles per hour. The weather was nice, the "big sky" country scenery was great, and there was no other traffic. The route was pretty much in one direction. In that part of Montana, there are no mountains to dodge, and very few streams to cross. In this "dream" state, I approached a small narrow bridge over a dry stream bed. I automatically steered the tractor for the center of the bridge to have ample room. Within a few feet of the bridge side supporting structure I realized the rear towed cultivator was wider than the bridge! The test crew following in the pickup saw what was about to happen but had no way of warning me. With much haste, I was able to brake the tractor and stop before causing an embarrassing failure report to be sent back to Waterloo.

The Engineer-in-Charge of the Nebraska Tractor Testing Laboratory in 1960 was Lester F. Larsen. He is known for his technical expertise and integrity. His dedication and belief in the tractor testing program prompted him to help leaders in other countries establish similar programs. In later years, changes have been made to permit global manufacturers to test in one country and gain test acceptance in most other countries.

"The evaluation and field test crew were at the field test site by 8:00 a.m., carried a lunch, and didn't leave until 5:30 to 6:00 p.m. A maxim often stated was that an engineer wasn't a true Deere engineer until he had green blood and had been to Laredo, Texas. The evaluation crews were usually there for 4-8 weeks extended residence.

"Customarily design engineers of both tractor components and implements were invited to visit the site and work on the many mutual problems. Field evaluation could continue for almost 12 months of the year, including time in the upper midwest when the weather was favorable."

Nebraska Tests

Any discussion of tractor testing should include comments about the Nebraska Tests. These tests were authorized in 1919 by the Nebraska state legislature. The principal provisions were:

- No tractor could be legally sold in Nebraska without a permit.
- The permit was issued by the State Railway Commission after completion of satisfactory tests.
- The test results were compared to the manufacturer's published claims.
- The tests were performed by the University of Nebraska Agricultural Engineering Department and certified by a Board of Engineers.
- Provision was made for a temporary permit until the tractor could be tested and the test reported.

The 3010 and 4010 tests were conducted and reported in 1960. Their test numbers follow:

759 — 4010 gasoline
760 — 4010 LPG
761 — 4010 diesel
762 — 3010 diesel
763 — 3010 gasoline
764 — 3010 LPG

This 3010 gasoline tractor was used in Nebraska Test Number 763.

This 4010 diesel tractor was used in Nebraska Test Number 761.

From the standpoint of the Product Engineering Department several goals must be met in testing. These goals were:

- That the tractor model being tested be representative in performance, parts, and construction.
- That the advertised power ratings be met or exceeded.
- That no significant failures occur during the tests.

Tractor preparation (there was usually a backup) involved many items. One was arranging to follow the tractor assembly and test in the factory. Another was the engineering break-in and testing to assure performance. Nebraska Tractor Test performance was expected to represent what a customer's well broken-in tractor was potentially able to achieve under the same standard test conditions. In the tractor preparation some latitude was permissible in tractor tire lug height (or wear) and the rubber hardness. However, the ambient temperature considerably affected engine and tire performance. Often the Nebraska Laboratory had to begin testing in the summer before daybreak in order to complete some work before the heat of the day. Certain ambient temperature objectives were met by air circulation control in the laboratory. Testing was seasonal in that no tests were conducted or scheduled during some winter months. The test conditions are stipulated in ASAE Standard S209.5 (and SAE J708).

Summary

Fortunately, Deere & Company believed in thorough experimental testing and recognized this as an important part of any experimental program expense. The roughly four years of testing enabled these entirely new tractors to achieve customer acceptance.

The 3010 and 4010 were in production approximately three years before they were replaced by the 3020 and 4020 models. The changes introduced by the later models were principally:

- Optional full Power Shift Transmission (PST).
- Power increases (this enabled the 4020 with PST to perform equal or better than the superseded 4010).
- Heavier rear axle housing, axle and optional rear wheels. This better adapted the tractor to wheatland operations.

The latter change was the only significant one that related to a field test inadequacy of the OX and OY tractor program.

The drawbar performance of a competitive tractor was tested in 1968 on this Nebraska concrete track.

Service

Fred Hileman: Early in our program I sent Service Representative J. W. 'Tim' Riley to Engineering. I told him to get as familiar with the product as he could. One item he found that gave us much concern was that to re-ring the engine it was necessary to pull the cylinder sleeves. The connecting rod would not go through the cylinder bore. At first, Engineering did nothing, but we continued to insist. Even Plant Manager Harley Waldon supported our viewpoint. Eventually a compromise was reached and a solution was made. We were looking for that type of problem.

Harley Waldon, Waterloo General Manager, 1956-1967

The Service Department activities are well described in *Service — The Complete Story*, the John Deere Service Bulletin No. 294, dated October 1960. The following is partially excerpted from that publication. The accompanying sidebars are from a conversation with Fred Hileman, Deere Waterloo Factory Service Manager through the time period covered by this book.

Fred Hileman, Deere Waterloo Factory Service Manager from 1938 to 1963.

The service story begins with the decision to produce the new line of tractors. Management realized that the over-all program would not be complete without a carefully planned, well-organized service activity. While engineers were busy designing and testing, members of the Service Department were busy planning and developing their areas. The four essential, closely related areas were: service school courses, service publications, special tools, and parts program. These big jobs involved many people. Of course, secrecy of the entire program was necessary.

Hands-on service training was provided for engine disassembly, inspection, and rebuild.

Parts Specialists

The parts drawings were studied with particular attention for the material being used. As the experimental tractors were being built, the individual assemblies and sub-assemblies were examined. The assembly procedures were observed to determine whether to furnish individual parts or the complete assembly for replacement.

The parts specialists were present when the test tractors were torn down. They studied the wear patterns and discussed them with the engineers. All factors that affected the service life of the part were considered.

With the design and testing continuing, the list of parts was revised to agree with the latest engineering changes. Eventually, a list of all the required service parts was compiled. Orders for service parts were then issued, stored when received, and prepared for shipment. Dealer kits were assembled. The goal was to provide each dealer with enough essential parts to adequately perform initial or minor service on the tractors first received. Additional parts inventory was provided for branch house storage. Quantities varied by branches as determined by the number of tractors intended to be shipped to them. An average of 42,000 pounds of parts was shipped to each larger branch, and 20,000 pounds average to each smaller branch. All shipments were made 30 days before the dealers received their first tractor.

Hileman: I had to go to Africa. So, I prepared Tim Riley how to reply should the marketing people from Deere and Company inquire about our service plans. On training, he was to say we could give a four-week course to branch house service persons to prepare them as instructors of dealer service persons. Upon my return, Harley Waldon said, "A meeting was called in Moline to review your service program plans. Tim presented the plans. They are quite concerned about your preparation for this." I reviewed with Tim what happened at the meeting. He responded that he had faithfully described the four-week program as suggested. We concluded that if we had said, 200 hours instead of four weeks, those college types would have a better appreciation. I went down to Moline and gave the same presentation, except I told them it was a 200-hour course. That was the last I heard about it.

Training Servicemen

The goal was to organize a factory training program or service school. The attendees were to be branch house service representatives. They, in turn, were to conduct training schools in their branches for the dealer service personnel.

The branch house people were trained in a school at the Tractor Works. Each branch divided their service staff into three separate groups. Each group attended classes on four-week intervals. When these classes were completed, eight factory instructors had qualified more than 150 branch house service men.

Preparation for the classes was done as the experimental tractors were inspected. Assigned time with an experimental tractor helped develop the story to be taught. The engineers were queried why given design config-

This service training lecture focused on engine disassembly.

Hileman: I don't know how many times marketing people would call and tell me they wanted the publication materials available with the first tractors. They didn't want something coming any later.

This service training class was learning about 3010 and 4010 tractor hydraulic systems.

urations were chosen. With each major assembly, a teaching plan began to develop. Gradually, these courses were developed: engine disassembly, transmission and final drives, hydraulic systems, fuel system, electrical system and tune-up, and separation of engine from transmission.

The class sessions opened with the theory of operation, then the proper methods of disassembly were taught. These sessions were followed by parts inspection for wear, and then correct assembly practice. Lastly, each participant would experience disassembling and assembling using the proper special tools.

It took about two years to accumulate the information for the school. Instructors with very limited knowledge of the two-cylinder line were trained to replace the regular factory service men.

The courses were given in the 7th floor of the parts warehouse building (demolished in 1995). Equipping the classrooms with partitions, light, heat, water, workbenches, and tools was done. Various training aids, wall charts, and service manuals were provided.

Hileman: Once Harley Waldon said, "'You're crazy to staff your factory training schools with men who know nothing about the two-cylinder tractors." I said, "There will to be too much time spent on the question 'you made the two cylinders that way and why have the new ones been made this way?' I'm sure the branch people will ask, and I want as clean a break as possible. I've given this a lot of thought, and I think this is the way to do it." He said, "Okay."

These men were listening to a service training lecture on 3010 and 4010 tractor transmission and final drives.

Service Publications

The service publications are an important part of the total tractor service program. Management stipulated early on that the manuals and parts catalogs would be available with the first new tractors and implements. All the information and photographs had to be new because the tractors were so different from previous models. This added to the burden of being on time.

Early consideration was given to having an outside firm prepare the publications. Before this could happen, however, it was decided to keep the work in-house and under control. The services of the Moline general publications group were not to be used.

Staffing the work, yet taking care of the needs of the service problems of the production tractors, required adding new persons for a new service department. Hileman commented:

> *"We placed a series of want-ads in area newspapers. And we acquired people. One most fortunate hiree was Marv Frey. He could make cutaway drawings, like the whole 4010 tractor, from just a stack of engineering drawings"*

A new building was constructed at the Product Engineering Center to house the writers, artists, tractor teardown, and photographic needs. The building, fronting on Ridgeway Avenue, was later used by the farm staff, before being demolished in 1995.

Revisions were made to incorporate the latest engineering designs right up to the release to the printers. More than 6,000 new illustrations were part of the manuals. More than 100 tons of paper were used to print the manuals for the first tractors. A larger 8-1/2- by 11-inch format was used which permitted larger illustrations. A new plastic shipping envelope was developed. This same envelope could be used as a protective jacket for later storage by the customer.

Hileman: One morning Morris Fraher, Vice President Tractor Production, called. He said, "We've had a meeting. It has just been decided that we will do our own writing, and preparation of all service publications." And he says, "You are it." I said, "Well, we're partly prepared. Can we use the Moline Service Publication staff that has been doing our two-cylinder materials?" He said, "They must not know anything about this." I said, "We have to have camera people, I have no writers, and no artists that can make the drawings." He said (in his usual direct manner), "Get them." So I went in to Harley Waldon and said, "I want you to know I'm going to be spending some money." His reply, "Well, if Morris Fraher said do it, do it."

This cutaway drawing shows the inner workings of the 4010 tractor.

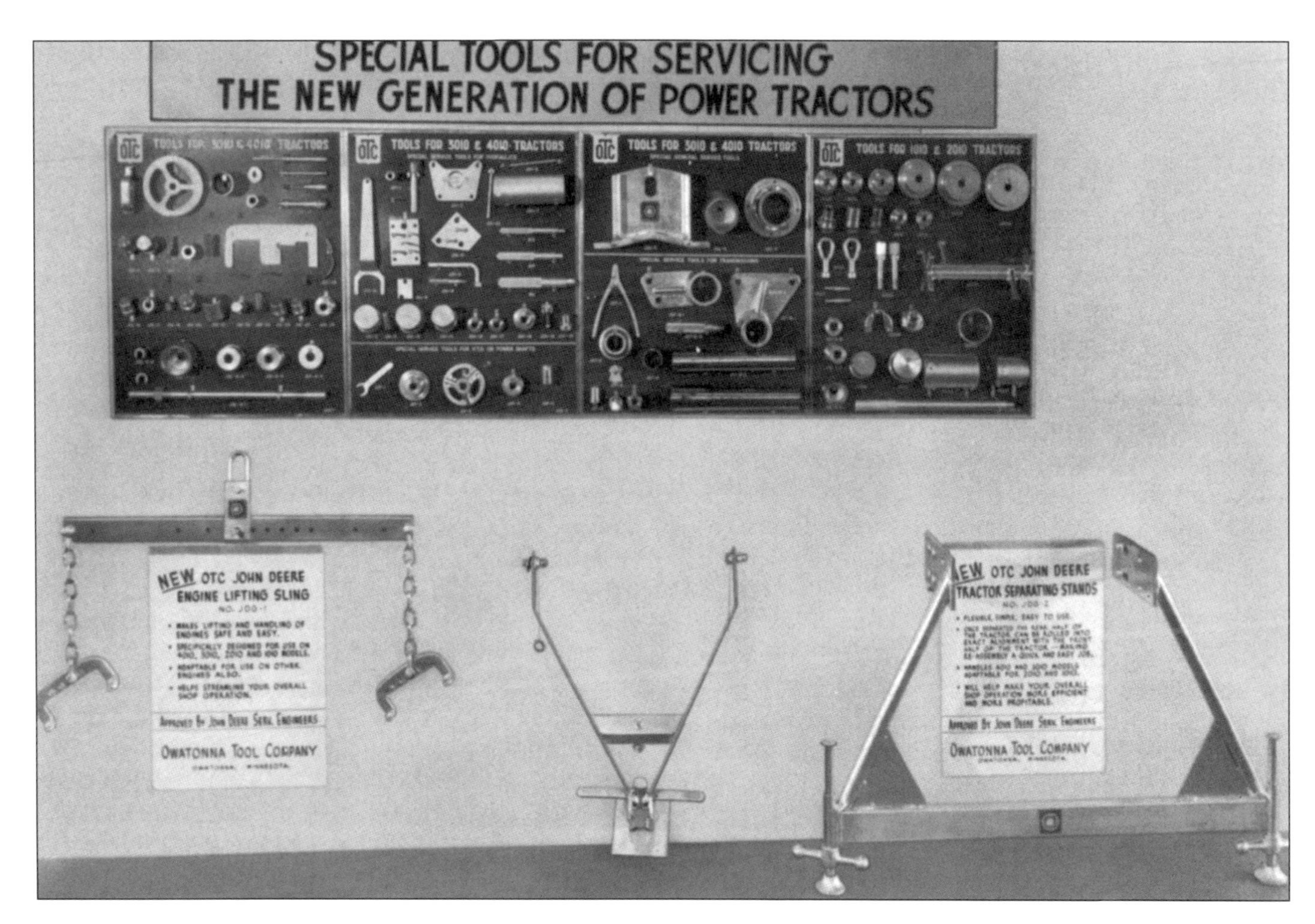

This display shows the special tools recommended for the 3010 and 4010 tractors.

Special Tools

Certain special tools, some similar to what the engineers used, were developed to permit better disassembly and assembly practices. Still other tools were similar to ones developed by the engineering shop personnel. The need for even more tools was identified by the experienced factory service persons. The special tools were made by the Owatonna Tool Company of Owatonna, Minnesota.

Hileman: "We had an agreement with the branches that no dealer would get his tractor until he got his tools. We had everything boxed to go out. We had some problems with part of the St. Louis branch where they said they would only sell the 3010. They wouldn't take the whole tools package as we had set it up. This meant we had to warehouse 4010 tools until 4010 tractors were sold in those territories."

Hileman: Morris Fraher recognized that these new people needed a place to work. He suggested some remote corner of the plant. Then he suggested a building be provided on what became the Engine Works property. On both suggestions I pointed out that the frequent transport of experimental tractors for our use would cause special security problems. He said, "Fred, I'll be up at Waterloo next week, and we'll discuss location some more." When he came, we again reviewed his proposals. Then I said, "If you're going to build a new building for me, why not just build it across the road on PEC property?" Well, he still didn't think that was where it should be. A couple of days later I got a phone call from him. He said, "Fred, it's been settled. We'll put the building across the street on the PEC property."

Manufacturing and Reliability

Placing the manufacturing and reliability topics near the end of this book may be in proper chronological order. Of course, the factory and reliability arenas need a product design before their functions can begin. However, in this case being last does not mean being least. All parties needed to be consulted almost from the "first lines on the drawing board." And no new product would satisfy the customer without being dependable and without being made and assembled with the lowest possible costs.

Manufacturing

From the beginning there was no question that the new tractors would be built in Waterloo. The considerable investment in buildings and trained, skilled workmen must have weighed significantly in considering Waterloo. Manufacturing includes the casting of all gray iron parts, machining iron and steel parts, and forming sheet metal parts. Also included is the sub-assembly, testing, and complete tractor assembly processes. Certain specialty items such as bearings, tires and rims, and electrical components were purchased.

Early in the development, a Manufacturing Engineering team disassembled competitive tractors and estimated their manufacturing cost. This estimate gave managers a guide for the new tractor's costs. Throughout the program some persons from the Manufacturing Engineering team spent part of their working time consulting with the Product Engineering personnel. Other Manufacturing Engineering persons with specialized skills were available for consulting upon request.

Hot iron was being poured on the foundry cupola line.

At this point let me clarify the meaning of the term "factory engineer." To a Product Engineering person, the term usually meant a member of Manufacturing Engineering. To the customer and dealer the term usually meant a Product Engineering person(s) who designed or tested the tractor at some time. In the latter case if the situation of contact was pleasant, the term was complimentary, elevating the individual to an "ivory tower" respect. On the contrary, the situation could be one of tractor failures, with attendant down time and crop loss. Then the term could be applied with a derogatory "they should have known better" meaning.

Manufacturing Engineering had very important decisions to make to launch the new product:

- New major machines had to be acquired and installed. In this category, the new cylinder block, crankshaft, connecting rod, and a host of other parts required new machines.
- The disposition of older machines that made obsolete parts. Many machines, as suggested earlier, had reached the useful end of their productive life. Many others were serviceable, but not adaptable to the new products. Some were retained for making service parts for older tractors. Much of the equipment used to make the 730 diesel tractor was shipped to Argentina where that two-cylinder tractor continued to be made and sold. Some factory machines could be adapted to certain new tractor parts. For example, an automatic screw machine can make a wide assortment of parts from bar steel.

The Waterloo factory contained several multi-story buildings. Many parts were to be machined on upper floors. The connecting rods were one of those parts. Installing several machines weighing approximately 25 tons was required on the 6th floor of one building to build the connecting rods. The elevators used for normal parts transport could not accommodate them. A crane was brought in, a section of roof was removed, and the machines were lifted to their position.

Major aspects of the factory preparation conducted by Manufacturing Engineering included:

- Assuring the availability of current tractors to meet the continuing customer needs of the Marketing Department. This procedure required stockpiling complete parts and partially assembled tractors, without sheet metal, for nearly a year to make time for factory rearrangement and the installation of new machines.
- Determining the factory arrangement to produce the new tractors while continuing to supply the repair parts for all discontinued tractors.
- Committing to a production start date, deadlines were then given to Product Engineering for release of various broad parts categories. Major new machines that required more lead time for procurement were of concern, along with the time required to set up to make trial parts.
- Shifting personnel to new assignments after Product Engineering released the tractor build specifications and the increasing quantities of new parts. In most cases existing machines had to be shut down, moved, and retooled to provide efficient routing for the parts.
- Making the factory machines and their tooling available. Product Engineering needed to make design changes as a result of their continued testing. Thus, the work of each group was overlapping in time. Consequently, as the factory equipment installation proceeded, many conferences were held to accommodate the product design changes.

This machine drilled the engine main bearing oil passages.

This photo shows the tractor assembly line used in the 1960s.

These tractors have reached the end of the assembly line. There is a thrill watching the previous collection of parts and pieces change at startup into a noisy, throbbing machine with a seeming "life" of its own.

Reliability

Assuring dependability was the function of the Quality Control Department. They set up monitoring plans to document the statistical variation of dimensions and surface finishes on all parts. This procedure assured that the parts would assemble properly and give the expected operation life.

This gentleman was measuring the smoothness of the engine crankshaft bearing surfaces.

As problems arose in obtaining the necessary dimensions, finishes, or assembly procedure, Quality Control focused on corrective action. Quality Control was also responsible for assuring that the two-cylinder tractor assemblies in inventory deteriorated minimally. Also, that they were properly returned to the factory and finished as needed.

Product Engineering Follow-up

Assistance to Quality Control, Manufacturing and Service departments was given by the Current Products Engineering Division. This group continued to occupy the old Product Engineering offices at the factory (later occupied by Manufacturing Engineering). They did not move to the Product Engineering Center until it was expanded for them in 1964. With parts released to the factory by the New Products Engineering group, the new tractor's product design continued with a small Current Products group. This latter group was located conveniently at the factory for consultations.

Timely delivery of certain machines could not be obtained from United States sources and had to be purchased overseas. On one occasion the machine was shipped air freight from Germany. While unloading in Chicago, the machine fell from the forklift as it was lifted from the plane. Considerable scrambling occurred to repair and get the machine installed in Waterloo to produce the required parts.

This 1,000-ton capacity press was used for making the hoods.

The Current Products Engineering responsibility was aided by the transfer of personnel from New Products, who were familiar with the design, to be closer to the factory. Personnel exchanged from Current Products Engineering were assigned to various new programs at the Engineering Center.

The Effective Result

The effectiveness of the total effort by all engineering and factory personnel was forcefully demonstrated by Fred Hileman, then the Factory Service Manager. His statement was given in 1961 (after one year of 3010 and 4010 production) at a marketing tent show at the Waterloo airport.

When it was Hileman's turn at the podium, he walked up carrying a small black satchel. He placed it on a nearby table, then announced, "In this bag are one of each of all the replacement parts needed for 4010 and 3010 tractors this past year, except one. And I want to tell you about that one!" He continued saying that full load 1st gear testing of a production transmission at Product Engineering revealed a structural weakness in the differential gear carrier housing. He spoke further of how quickly a design correction was made, and revised parts were produced and installed in both production and tractors already built. Then he told about the other items from the satchel.

The culmination of all this effort was for one purpose — the new 3010 and 4010. In the early years of the tractor industry, many wholly new models went into production directly from the drawing board. By the 1960s this industry was a mature market. New entries were expected to have equal or superior features, and they needed to have dependability that was also equal or better. By all measures these new tractors received high marks.

These new tractors started a long standing production run (including the later-derived models). They continued Deere's dominance of the country's agricultural tractor production volume, and are a credit to all who were involved. The marshaling of all associated persons reveals the commitment, the dedication, and the "esprit de corps" that was instilled by the various leaders. The prospects are unlikely that an event of this significance will occur again, given the cyclical, mature nature of the agricultural tractor business.

The Waterloo product engineering office was a very busy place in the 1950s.

Fred Hileman, Factory Service Manager, related how this presentation developed: "One day Harley Waldon came into my office. He said, 'Fred, the Marketing Department down at Moline have just talked to me about a coming meeting of branch managers and board members. They want to talk about this tractor and the troubles we've had. I don't know anybody that knows more about the troubles you've had than you. Would you get ready to talk to them?' Well, I got out all the bulletins and studied them. I asked some of my people to get all the new parts. And I put them in the satchel I had. Except the differential.

"I took the satchel with me in Harley's car to the meeting. Once Harley lifted the bag, and said, 'Hileman, what have you got in here?' I said, 'I'm taking a set of tools down to Moline and I'm going to pull a job.' Well, the job I pulled was so successful that I spent all next winter travelling to branches for repeat presentations.

"Now there's another story about the differential. Among the tractors already shipped, we found one of the tractors had been purchased by International Harvester. The dealer was supposed to replace the differential. Instead, he loaded the tractor on his truck and brought it to Waterloo. Our Engineering Department made the replacement, but in the process found tracks that IH had left in their disassembly examination."

The Public Announcement

When tractors are rolling off the assembly line, announcement to the media, dealers, and customers is timely. This announcement requires months of planning by the Marketing Division. Seldom is a strategy used in which the new product is unveiled to the public, dealers, and customers early in the prototype and experimental phases. Most new products replace existing models in the line. The existing models are then discounted in price after the new model is announced. Despite the discount, there are many customers that want to buy the latest and greatest. Therefore, sales may be slowed of the remaining old model. The company needs to continue selling the old model at the highest price possible to help pay for the expenses of bringing in the new product.

In the months prior to the unveiling of the new tractor line in Dallas, Deere personnel were making sure that all the minute details were covered. They rented a building in Moline and reproduced in miniature the Dallas Coliseum and Cotton Bowl parking area. Toy tractors, models of people carved out of wood, and replicas of tents and other items (all to scale) were used in determining how the Coliseum and parking lot display would be set up. (The Dallas Times Herald)

One ploy that was used to conceal the shipment of the early tractors was a kraft paper cover. These covers concealed the hood, engine, frame and seat from view. They looked like an extra protection from the elements as tractors were shipped on open trucks or railroad cars. The covers were used after the Dallas dealer announcement, but before public announcements. During this period each dealer was supplied with a few tractors of each model.

Security

Public announcements of new products are carefully timed. Precautions are taken prior to the announcement to limit the knowledge of the new product only to those who need to know. The 3010 and 4010 tractor programs had the tightest security guidelines ever exhibited at Deere in Waterloo. This measure of security ("need to know"), applied primarily to company personnel and supplier representatives. Not all company personnel were qualified as "needing to know" to perform their own assigned responsibilities. Some of the security precautions taken with test tractors and test sites have been described in previous sections, but other areas of communication needed security.

A list of qualified persons was maintained, and it was available to some managers and the lobby receptionist. The receptionist monitored the comings and goings of the listed outside persons. She was told of additions to the list and made appropriate revisions. She also advised anybody at Product Engineering Center (PEC) of an individual's security clearance status. Not all outside (of PEC) business persons were given complete disclosure of the product.

Supplier representatives may have only seen a layout of their particular interest area. If their responsibility required their presence in the experimental shop, they may have seen an experimental tractor with their product installed. They also may have seen other experimental tractors with a competitive supplier's product. It behooved them not to ask questions beyond the need of properly applying their product. As the experimental design progressed, there always was the need to add new names from within the company and among supplier representatives. New persons were made aware of what was expected of them and that Deere was monitoring the security.

A clearance procedure required that each new person be made aware of the security concern about confidential information. This procedure was performed by middle or upper level designated managers. The new person was informed that good business ethic was expected of them concerning disclosure of what they would see or hear.

Storage and handling of confidential papers was discussed. The extent of discussion with others in their company was described. Similar guidelines were used with Deere & Company persons from outside the Product Engineering Center, including from other factories. Once cleared, their name was added to the security name list.

Except for a few slips, the security measures were generally successful. As a result, most of the dealers assembled at the public announcement in Dallas, Texas, were surprised. They had no previous inkling of the magnitude of change occurring in the tractor line.

William Hewitt, president of Deere & Co., opened the Dallas with a welcoming address:

"These new machines have been designed from the ground up to meet the needs of your farmer and industrial customers.

"Product research, design, and development have been fundamenttal to our progress. Six years ago at Waterloo Iowa, we created the most modern tractor research center in the industry. Staffed with skilled engineers and equipped with complete testng facilities, this center was given the job of creating an entirely new line of tractors tailored to the exacting needs of today's and tomorrow's agriculture.

"With tractors so new, so different, so completely attuned to the need s of the market, we have chosen what we believe to be an approprite slogan: The New Generation of Power."

Dallas Day — August 30, 1960

Tuesday, August 30, 1960, is a day that many people in the tractor industry will never forget. Deere & Company brought more than 6,000 people to Dallas for the unveiling of the new tractor line. Why Dallas? Most previous company tractor announcements were made at one of the factories, or at the main office in Moline. Dallas represented a special destination — one that would appeal to more of the dealers.

The event was planned to the last meticulous detail. Some examples of this precise planning included guest accommodations and transportation. Five thousand Deere dealer family were moved in and out of Dallas in history's largest commercial airlift, plus 1,000 additional guests were transported by other methods. Hotel and motel rooms were all awaiting the guests when they arrived on August 29. The next morning the entire

The Dallas Coliseum on August 30, 1960, welcomed the Deere dealers with a Texas-sized JOHN DEERE sign. Note the buses in the foreground that were used to transport the guests.

A marching band proceeded the entrance of the new tractors. Lyle Cherry, General Sales Manager of the Waterloo Tractor Works, stood over the entrance and introduced the tractors.

contingent was moved from hotels/motels to the Dallas Memorial Auditorium by a fleet of buses. That evening they were moved to the Texas State Fairgrounds by the same fleet of buses. Transporting 6,000 people was no small task.

Deere & Company officials made elaborate arrangements to provide medical aid for their guests. This plan included contacting medical specialists of all kinds, including an allergy specialist. But apparently Deere dealers were a healthy lot. Only two incidents were reported: one man suffered heat exhaustion and a newspaper man broke his bridge while eating sweet corn at the barbeque. (The Waterloo-Cedar Falls Courier)

A Texas-Sized Welcome

The largest welcome sign in the history of Texas — where everything is always the world's largest — was wrapped around two sides of the Memorial Auditorium to greet the guests. Measuring 234 feet wide and 25 feet high, the nine-panel yellow canvas sign spelled out the name JOHN DEERE in huge, 18-foot green letters. The event opened on August 30 with a preview of the entire line which was enthusiastically received by the dealers. This display was held in an adjacent parking lot of the Cotton Bowl.

A special unveiling, at 12 noon, occurred in the prestigious downtown Neiman-Marcus department store. The store's president, Stanley Marcus, removed a curtain concealing a 3010 diesel row-crop tractor. Befitting its location in the famous jewelry department, the tractor was trimmed with diamonds spelling the maker's name: John Deere. William Hewitt, Deere's president, was also present at this event.

The festivities continued that night at the Texas State Fair Grounds where all 6,000 guests were treated to an old-fashioned, Texas barbeque. They polished off 5,000 pounds of beef, 1,800 pounds of ribs, 4,200 chickens, 300 gallons of ranch beans, 2,500 pounds of potato salad, 6,000 ears of corn, 1,500 pounds of Texas cole slaw, 15,000 sourdough biscuits, 7,500 apple turnovers, 300 pounds of butter, 10 barrels of iced tea and 200 gallons of hot coffee.

The livestock section of the fair grounds was scrubbed to spotless cleanliness to serve as the picnic grounds and 1,000 tables were set up in the stalls that would hold prize cattle at a livestock exhibition the following week.

The evening program included speeches by Hewitt and C.R. Carlson, vice president of marketing, and others. Guests saw a lot of farm equipment, but the programs were frequently interspersed with lighter entertainment. At the Memorial Auditorium, movies and television "in the round" were used to portray the Deere story and show the new tractors for the first time in color. Another feature of the morning show was a troupe of 32 ice skaters led by Mike Kirby, former skating partner of Sonja Heine. Al Hirt's Dixieland band was imported from New Orleans for the evening entertainment. A huge fireworks display in the Cotton Bowl was the climax of the days events. It featured, among many other spectacular pieces, life-size tractors in green and yellow with accompanying displays giving the various model numbers.

William Hewitt, Deere & Company president, mingled with the crowd at the fairground like any good farm equipment salesman would. His final words, in a brief talk at the Cotton Bowl, while rockets were still going off overhead were "I hope we can do this again some time."

In many ways the Dallas meeting was unique in the history of American commerce and industry. It was the first time that Deere & Company, or any other farm equipment company, brought all dealers together at one place at the same time. It is also believed to be the first time in history that a farm implement company had put together farm equipment in one mammoth exhibit worth more than $2 million. Deere personnel who were with the company, though not privileged to be at Dallas at that exciting time, still remember the impact left by this significant marketing announcement.

A highly secretive sales meeting took place in Waterloo a few weeks prior to the Dallas event. Some Deere sales people were given a preview of the new tractors at an exhibit set up on an isolated part of the Municipal airport property. Despite the secrecy, a plane bearing a load of engineers from a competing company circled above, apparently trying to give the engineers a close look at the new Deere equipment. (The Waterloo-Cedar Falls Courier)

The tractors were on display at the Cotton Bowl parking lot at the Texas State Fairgrounds. Dealers were given the opportunity to see the tractors first hand after the introduction at the coliseum.

The Final Word - The Tractors

The measure of success of any effort is in the results. The total 3010 and 4010 tractors that were built and sold over the next three or four years were about 45,000 and 40,000, respectively. They were instrumental in raising Deere & Company's total agricultural sales to number one in the United States. The industry's informal marketing cry changed from "get IH" to "get Deere."

Further, the Waterloo-built successor row-crop models have continued Deere's major share of the U.S. tractor market in their power sizes. There were several days in 1965 that the Waterloo assembly line general foreman, Bill Horan, spoke of assembling over 100 4020 tractors per day. The total of more than 175,000 4020 tractors made it the most widely sold model ever built by the company. The popularity of the tractors, plus the lack of need for major updating, resulted in the continuance of the 3020 and 4020 models for eight years.

Many factors deserve credit when a successful new product is created, introduced, and continues a long production life. In this writer's opinion the most important factor of the 3010 and 4010 program was the people. It began with the support and direction given by Deere's chairmen and the board of directors. The managers at Waterloo and Moline provided the guidance the product engineering staff needed. The lead project engineers were given the freedom to be innovative in meeting the challenging objectives set before them. And lastly, the unstinting concern and cooperation of all who were part of the team working on the program.

After years of hard work, the tractors were finally ready to make their debut. They were shipped by truck and train.

This 3010 diesel standard tractor worked in rice country (note the cane and rice tires).

A preproduction 3010 gasoline tricycle tractor and a PTO-driven 30 combine were harvesting soybeans.

This 3010 row-crop diesel tractor was operating a 406 lister planter.

A 3010 gasoline row-crop tractor with an adjustable front axle was pulling a 323-W baler (note the power-adjusted rear wheels.)

This 3010 gasoline row-crop tractor was pulling a plow on rolling terrain (note the front rockshaft).

This 3010 diesel special row-crop utility was rare. It was for export only and unusual characteristics include the small fenders and the steel seat.

A full cab was available for this 3010 diesel industrial wheel tractor.

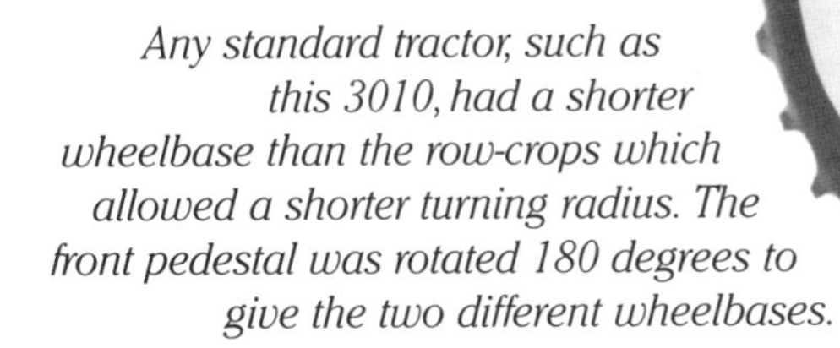

Any standard tractor, such as this 3010, had a shorter wheelbase than the row-crops which allowed a shorter turning radius. The front pedestal was rotated 180 degrees to give the two different wheelbases.

This 3010 grove and orchard tractor was at work amongst the blossoming trees.

To provide a sleek profile and to protect the operator from tree branches, the orchard version of the 3010 had a lowered operator's station.

The LP-gas tractors were distinguishable by the tank protruding from the front hood. Pictured here are two 4010 LP row-crop tractors, the top tractor was pulling a rear-mounted subsoiler and the bottom tractor was pulling an integral plow.

This one-of-a-kind 4010 gasoline Hi-Crop tractor was never produced.

The majority of the 4010 Hi-Crop tractors were diesels, only a small number were LP gas. This diesel tractor pulls a Model 67 tool carrier and border disk.

Another view of the 4010 Hi-Crop gasoline tractor which never made it to production. Factory records indicate that it was "scrapped" and rebuilt into a diesel (it also had a new serial number).

The 4010 diesel row-crop tractor and the 730 row-crop tractor stand at the ready. This photo provides a good comparison of the two models.

This John Deere display was at the 50th Waterloo Dairy Cattle Congress in September 1962 — which also was the 125th anniversary of John Deere.

This 4010 gasoline row-crop tractor was running a pto-driven 10A hammer mill (note the wide front end on the tractor).

This 4010 row-crop tractor was pulling a mounted beet harvester.

This 4010 diesel row-crop tractor had a single front wheel and was operating a one-row high-drum cotton picker — an impressive and unusual sight.

This 4010 diesel standard was outfitted with an early factory cab. It was operating a 96 combine in windrowed wheat.

Diesel engines were popular and accounted for over 80% of sales for all types of 4010 tractors. A 4010 diesel standard pulled two CC-A field cultivators in this photo.

LP-gas engines outsold gasoline engines with over 10% of sales. A 4010 LP row-crop tractor operated a 6-row 870 series integral lister-planter.

This 4010 diesel standard was levelling the land with for irrigation with a 940 Landshaper.

This cutaway 4010 diesel was functional (note the driving electric motor below the engine oil pan) and was on display at the 1962 Waterloo Dairy Cattle Congress.

The 3010 diesel row-crop and 4010 diesel row-crop were displayed at the Moline Administrative Center.

The 4010 diesel special standard was for export only. Like the 3010 diesel special standard, it had a steel seat and small fenders.

The Numbers

Nebraska Tractor Tests

Model	3010	4010
Year	1960	1960
Test No.	762	761
Max. PTO Hp	59.4	84.0
Drawbar:		
Max hp at 5.8 mph	54.5	
Max hp at 5.7 mph		73.6
Max. pull (lb)	6,323	7,002
Weight w/ballast (lb)	8,640	9,775
Fuel Use at:		
Max. PTO hp at std PTO rpm (hp-h/gal)	15.5	15.7
75% of pull at max. power (hp-h/gal)	13.0	12.4

3010 and 4010 Specifications

Model	3010	4010
Transmission	8 SR	8 SR
Engine		
Cylinders	4	6
Bore & Stroke	4.12 x 4.75	4.12 x 4.75
Disp. (cu. in.)	254	380
Rated rpm	2,200	2,200
Wheelbase (in)	90	96
Rear tires	13.6-38	16.8-34
Last list price	$5,362	$6,167

The 3010 was offered in more models than any other New Generation tractor built at Waterloo. It came in the row-crop, the standard, the row-crop utility, the industrial, the special row-crop utility, the Lanz standard, and the orchard.

The 4010 was offered as the row-crop, the standard, the hi-crop, the industrial, and the special standard.

Production of both the 3010 and 4010 ended in July 1963.

References

Broehl, Jr, Wayne. *John Deere's Company.* Doubleday: New York, 1984.

Buckingham, Earle. *Spur Gears: Design, Operation, and Production.* McGraw Hill: New York, 1928.

The Dallas Times Herald. Part D. August 30, 1960.

Fundamentals of Machine Operation. Deere & Co.

Gray, R.B. *Development of the Agricultural Tractor in the United States — Part 1.* USDA, 1954.

John Deere Service Bulletin, No. 294, October 1960.

Larsen, Lester. *Farm Tractors 1950-1975.* ASAE: St. Joseph, MI, 1981.

Morling, Roy. "Agricultural Tractor Hitches. Analysis of Design Requirements." ASAE Distinguished Lecture Series, No. 5, 1979.

Purcell, William F. H. "Industrial Design A Vital Ingredient," *Automotive Industries.* May 15, 1961.

Severin, William. "Deere Officials Proud of Dallas Reaction to Tractor," *Waterloo Daily Courier,* September 1, 1960, p. 31.